F

MÉTÉORITE

METEORITE

HOW STONES FROM OUTER SPACE
MADE OUR WORLD

TIM GREGORY

BASIC BOOKS
NEW YORK

Basic Books
Hachette Book Group
1290 Avenue of the Americas, New York, NY 10104
www.basicbooks.com

Printed in the United States of America

Originally published in 2020 by John Murray (Publishers) in Great Britain

First US Edition: October 2020

Published by Basic Books, an imprint of Perseus Books, LLC, a subsidiary of Hachette Book Group, Inc. The Basic Books name and logo is a trademark of the Hachette Book Group.

The Hachette Speakers Bureau provides a wide range of authors for speaking events. To find out more, go to www.hachettespeakersbureau.com or call (866) 376-6591.

The publisher is not responsible for websites (or their content) that are not owned by the publisher.

Internal artwork drawn by Rodney Paul.

Typeset in Bembo MT by Hewer Text UK Ltd, Edinburgh

Library of Congress Control Number: 2020940088

ISBNs: 978-1-5416-4761-9 (hardcover); 978-1-5416-4760-2 (ebook)

LSC-C

10 9 8 7 6 5 4 3 2 1

This book is dedicated to you.

All you that do behold my stone,
O; think how quickly I was gone;
death doth not always warning give
therefore be careful how you live.

This elegy is carved into a headstone in the Wold Newton churchyard, East Riding of Yorkshire, commemorating the life of John Shipley (1779–1829). Shipley was a ploughman of the nearby Wold Cottage estate. He almost became the first person in recorded history to be hit by a meteorite falling from the sky.

Contents

Meteorite Family Tree   xi

Prologue: Written in Rock   1
Stones from the Sky   7
Earthfall   35
From Gas to Dust; from Dust to Worlds   61
Spheres of Metal and Molten Stone   83
Cosmic Sediments   107
Drops of Fiery Rain   131
Stars Down a Microscope   153
Star-tar   183
Pieces of the Red Planet   211
Calamitous Tales   239
Epilogue: The Story Goes On   269

Appendix: Meteor Showers   273
Acknowledgements   277
Notes   279
Index   285

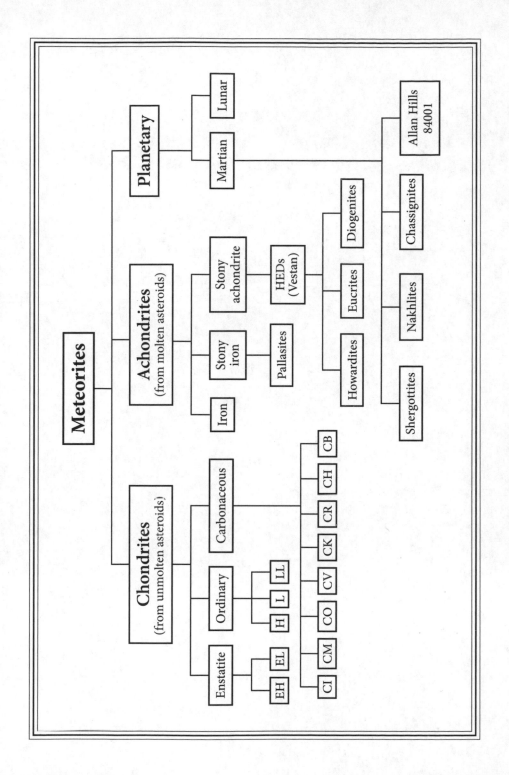

METEORITE

Prologue: Written in Rock

The oldest stories tell of fantastic beings and extraordinary events. Chariots drawing the Sun across the sky; great snakes devouring worlds; spirits singing the Universe into existence; Creation itself arising from the dismembered bodies of deities. Such stories were told by people who lived before the advent of the scientific method and so, although to our modern minds they may seem remote from reality, they provided a framework in which our ancestors could understand the world.

For most of human history, these stories passed from person to person through the spoken word. The mind was the only place where they could be stored. If a story or snippet of understanding was not spoken, heard, and then remembered by somebody else, it died along with the mind in which it sat. Only through the spoken word and commitment to memory could stories transcend a human lifetime.

It wasn't until some 125,000 years of human history had elapsed that stories and ideas were liberated from minds and rendered as objects in the real world. The oldest known human artwork (at the time of writing) was discovered in the Blombos Cave in South Africa and dates back some 75,000 years. It is an unassuming shard of stone a few centimetres across – easily small enough to clasp between thumb and forefinger – and embellished by a cross-hatch pattern drawn with an orange-red clay. The first human artworks were drawn using rocks as both paper and ink: rock was the first medium on which human thoughts were consciously recorded.

Recording stories and ideas in this way became a long-standing tradition, and as the slow march of human history progressed, the physical representations of thoughts became more sophisticated. Stories began to be written down. Thirty-five millennia after the cross-hatched pattern was drawn onto the stone in Blombos, our ancestors were drawing animals on the walls of the limestone caves of Maros in Indonesia. They drew these animals because they were important to them: they were the source of food that kept them and their fellow tribespeople alive. Their drawings show a deep respect. These animals would have been a major part of the story of the artists' world and life, and so they recorded that story permanently on the rocky walls of the caves.

The people living in those caves also left shadows of themselves on the rock. Inside the caves of Maros lie a dozen stencils of human hands. Their creators made them by pressing their hands against the cold cave walls and spraying them with a wet pigment, probably from their mouths. This simple act marked another great leap forward for our species' consciousness. The handprints are explicit attempts made by our ancestors to leave a trace of themselves upon the physical world, an ancient version of 'I was here'. They probably had a concept of a tangible future, a time when they would no longer walk the Earth, when the story of their world would continue without them. We see their small story today, 40,000 years later, by reading it on the rocky limestone walls.

Anything that is recognisable as *writing* didn't come along until about 5,000 years ago, by which time over ninety-seven per cent of our existence as a species (so far) had passed. Written words were first created by pressing symbols into clay tablets and chiselling them into tablets, and later by scratching ink onto parchment. Their purpose was elegantly simple: to allow thoughts and stories to be recorded in the physical world, precisely and accurately, and read later by somebody else. It is a means, other than the spoken word, by which the thoughts of one person can be placed into the mind of another, but it transcends the lifetime of the thought's original bearer. It is the closest humanity has ever got to mind reading, and the closest we'll ever come to communicating with the dead (though admittedly, the conversation is entirely in one direction). And it's all done via strange-looking shapes carved into stone tablets, scratched onto pieces of parchment with ink or, as in the case of this book, printed onto paper (or perhaps shining from an ebook reader).

Writing changes the way thoughts are recorded, and eliminates much scope for ambiguity and changes in meaning as they are passed from mind to mind. It also means that it is easier to catch up on things that have already been discovered; knowledge and ideas do not have to be acquired from scratch by each new generation, freeing up our time and headspace to make new discoveries and plumb the depths of new understanding.

Our acquisition of knowledge and understanding of the world accelerated as a result of the long tradition of recording stories, and it all began by recording them on rocks.

Parallel tales

There is another story written in rock, but this one wasn't written by us. It was written by Nature, and it begins further back in time than our own human story. Much further back. And it unfolds over a timescale so vast that it is impossible for the human mind to fully comprehend.

It began about four and a half billion years ago. This is time on a scale that dwarfs the 200,000 years during which humans have walked the planet, and it has an appropriately evocative name: 'deep time', or 'geological time'. This is the reel on which Earth's story plays out. It unravels on timescales where continents move, sea floors are raised to jagged mountain peaks, and millions of species come and go as they evolve before marching headlong into extinction.

The only way that the slightest sense of the immensity of geological time can be grasped is through analogy and metaphor. Imagine a twenty-four-hour geological day, where the entire four and a half billion years or so of Earth's history is squeezed down into twenty-four hours. On this timescale, the history of humanity would be just shy of four seconds long. The cross-hatch pattern from the Blombos Cave was drawn one and a half seconds ago. Writing was invented about one-tenth of one second ago, which, as it happens, is the time it takes to blink. Dinosaurs, often thought of as ancient life forms from the deepest recesses of the past, only evolved something like one hour and fifteen minutes ago, and their time on this planet came to an end after about fifty-five minutes. Trees have existed for a mere two hours. The land was little more than a bleak rocky desert until land plants evolved some two and a half hours ago, and the oceans were devoid of fish only half an hour before that.

The story of the Earth – its geological history – tells of the things that have happened since it formed to the present day. It has experienced unimaginable change during that time. Most of its story has been lost, much like our own human story, but plenty remains to be read if you know where to look and how to read it. Like the early chapters of the human story, this story is written in rock, too, and it can be read using the language of geology.

Like the pages of a book

Silt accumulating in the dark oceanic abyss; molten rock crystallising far below the Earth's surface; sand dunes rolling across ancient deserts. The circumstances under which a rock came into being can, more often than not, be deduced from its geological character. Each rock contains a single short story, but a sequence of rock contains a narrative that unfolded over a longer stretch of time. By scouring the globe and turning the pages of Earth's geological history, we have come to learn much about the story of our home planet.

But the history of the Earth, and by extension our own history, is but a sub-plot of a far grander story. It is a story that begins further back in deep time, and one that is set on a scale that far exceeds the planetary: it plays out on timescales that reach the astronomical and distances that reach the interstellar. It is the story of our Solar System.

We share our celestial neighbourhood with a multitude of planetary objects of which the Earth is just one: two giant gas planets, two giant ice planets, four rocky planets, hundreds of moons, and billions and billions of comets and asteroids. All of them orbit the same light-giving star at the centre of it all: the Sun. Our star. And while each planetary body has its own unique history and story to tell, they all, it transpires, share a common heritage. They belong to the same Solar System, and if we go far enough back in time, we find that their individual histories converge. The first few pages of their stories are identical.

The rocks of the Earth yield little information about the origin and the formation of our Solar System because there is a limit to how far back in time they can take us. For one, the Earth did not exist when the Solar System was forming. And throughout Earth's history, geological forces have continuously un-made and re-made the Earth's rocks, recycling them through the power of plate tectonics and wearing them down with weather. Early pages of Earth's storybook were torn out and lost, and many chapters have been overprinted.

Fortunately for us, there are rocks that *did* record the events that happened as our Solar System was assembling itself. Some of these rocks survive to this day and they are the most ancient objects in our cosmic neighbourhood. While there are still many questions that remain unanswered, by using the language of geology and the tools of science, we have come to understand the story these rocks have to tell of the early Solar System in brilliant detail. In uncovering this story we have discovered the origins of our Solar System, the planets, and ultimately, ourselves.

But these rocks did not come from the Earth. They fell upon it from the sky, and we call them meteorites.

I

STONES FROM THE SKY

As soon as the curious gaze of our distant ancestors turned skywards, they began to notice stars streaking across the Heavens. On any given cloudless night in any part of the world, provided sufficient patience is exercised, the celestial light show continues. These lights, called meteors (from the Greek μετέωρος (*meteoros*) which means 'raised from the ground'), have been a part of the human experience of life on Earth for as long as our species has existed. We have never known skies without them. They continue to capture imaginations the world over, and as they blaze across the sky as if they are 'shooting stars', we continue to cast our wishes upon them.

But meteors, and their bigger and more brilliant equivalents, fireballs, are not stars at all. They are the products of something so unexpected, so out of the ordinary, that it was not until around two centuries ago that their origin and significance was realised by the minds of modern science. Meteors and fireballs have their origins not in astronomy, but in geology. They are the products of rocks falling from outer space.

Racing between twenty and seventy kilometres per second, a typical meteor would travel the length of the United Kingdom in just shy of half a minute. Like the prow of a ship parting the water, falling stones part the air in the Earth's atmosphere as they press onwards across the sky, compressing the gas before their bow. The air caught in their path is pressed so strongly and so rapidly that its temperature instantly soars to thousands of degrees Celsius. It glows incandescent as a result.

The forward-facing surface of a falling stone is superheated to temperatures far exceeding that of freshly erupting lava from a volcano; the outer layers of the stone are slowly vaporised and unmade entirely. Most stones are utterly consumed as they fall, but a handful survive their fiery passage through the atmosphere, and make landfall on the Earth's surface. We call these survivors 'meteorites'.

The Field of the Sky

At twilight some 4,000 years ago in the Chaco region of what is today known as Argentina, people were awaiting the warm return of the Sun God from beneath the horizon when, the story goes, He came forth from the sky and fell to Earth. A brilliant light suddenly filled the half-lit sky and a clamorous noise spread through the air. He appeared before them as a large mass of iron, jet-black on the surface and silvery on the inside. Fires sprang all around Him. His coming to the Earth persisted in the collective memory and mythology of the native Chaco people for some 3,500 years until the arrival of Spanish conquistadores. Driven by the lust for gold, silver, and power, European colonisers claimed vast areas of South America for themselves in the sixteenth century. They discovered, no doubt much to their confusion, that the native people of the Chaco region wielded tools and weapons made from iron of an unusually high grade. There were local legends of a Sun God falling to the Earth as a large piece of iron in the distant past. In their arrogance, the colonisers dismissed these stories as mere rumour in the hope that the so-called piece of iron was in fact a large deposit of silver.

In 1576, the Spanish had the natives take them to the mass of metal. They were led along a network of well-trodden footpaths over a flat-lying plain void of waterways and rocks, but occasionally punctuated by hollows a few metres across. The area was named Piguem Nonraltá, which the Spanish translated as *Campo del Cielo* (Field of the Sky). Then they came upon it: a large, smooth-skinned piece of metal protruding out of the soft soil. It was almost two metres across, and weighed an estimated fourteen tonnes. Ignoring the fact that this was a spiritual place of worship for the native people, the Spanish broke pieces off the mass for investigation.

A blacksmith determined that the stone was made not of silver, but of high-quality iron. The colonisers mistakenly thought they had discovered an iron mine, and that this large piece, which became known as *Meson de Fierro* (Large Table of Iron), was just the tip. Many more pieces were found scattered over the area, but written reports of their discovery went largely unacknowledged for the next 300 years. The Meson de Fierro was visited for the last time in 1783 by another Spanish expedition, and its whereabouts have since been lost to history. It is unlikely the expedition of 1783 would have had the means to move such an enormous mass of iron: it was probably rolled over into a nearby hollow, and has since been buried by silt-laden floodwaters and shrouded from view.

The Campo del Cielo irons are not from an iron mine: they are meteorites, and we know with reasonable precision the date on which these enormous pieces of cosmic iron fell from the skies thanks to the bushfires they ignited. Vegetation and shrubbery were incinerated by the enormous fallen mass, and a vast area of the Field of the Sky was reduced to charcoal. By carefully measuring the isotopic fingerprint of the carbon in the charcoal, the date that the plants died – the date of the fire – can be measured: this in turn corresponds to the date of the meteorite fall, which turns out to be ~ 4,000 years ago, well within the lifetime of ancient stories and the ancestral memory of the native people.

Campo del Cielo is an unusually large meteorite. When it slammed into the Earth's atmosphere it was probably one enormous body but, as it fell towards the ground, the heat melted and stripped most of it away. Immense forces and pressure also played a part in reducing the size of the rock, tearing it into many smaller (albeit still huge) fragments. Over a dozen large pieces of this meteorite fall have been unearthed in the Field of the Sky, and there are probably more that lie yet undiscovered beneath the soil.

A piece named 'Optumpa', which is one metre across and weighs over half a ton, can be seen today on public display in the Natural History Museum in London. I saw it when I was nineteen on a school outing, my first visit to the museum. I remember saying to my A-level geology teacher, Mr Currie, 'I would love to work here one day.' As it turned out, I ended up completing parts of my PhD research there. I have had a particular fondness for Optumpa ever since.

Sky metal

In the arid region of central Iran lies the ancient city of Tepe Sialk. Archaeological excavation began early in the last century and hundreds of artefacts have since been uncovered: ancient architecture, sophisticated pottery, and solemn tombs. Amongst the items are a trio of small spheres or beads made of iron that date back ~ 6,300 years. What purpose they served is unknown. They caught the eye of archaeologists, however, not because of their shape, but because of what they were made from. These iron spheres pre-date the Iron Age by about 3,000 years. The technology used to smelt or process iron did not exist when they were crafted.

Modern analysis of the beads reveals their celestial origin. They are pieces of ductile meteoritic iron that were hammered into spheres by skilled craftsmen. In the Age of Stone, this metal – dense, shiny, malleable and cool to the touch – must have seemed strange to the people of Tepe Sialk. It is not known whether these pieces of iron were seen falling from the sky or were simply happened upon in the desert surrounding the city, but either way, the small iron spheres are one of the most ancient examples of iron being fashioned into objects.

There is something stirring about humanity's first contact with iron – a metal that so much of our modern civilisation is built upon – being iron of a celestial origin. It was the coming together of two stories that at first seem so separate but, as we will discover, are intimately entwined: outer space and humanity.

Fit for a king

Among the many treasures discovered in King Tutankhamun's tomb when it was unearthed in 1922 was a handsome iron dagger. With a handle of gold and a sheath to match, it had been placed inside the linen wrappings of the deceased pharaoh's mummified body before he was sealed inside his sarcophagus.

Lingering doubt about the spiritual significance placed on meteoritic iron was cast aside when the chemical character of this dagger was measured at the Egyptian Museum of Cairo. By irradiating the dagger with a focused beam of electrons, the object began to fluoresce in X-ray light, and by deciphering the 'colours' (wavelengths) of those X-rays, its chemical composition was revealed.

The dagger was found to be made of almost pure iron with a large serving of nickel: this iron, without a doubt, has a celestial origin.[1] Metal with this particular blend of elements does not form on Earth. Tutankhamun's dagger is made from an iron meteorite. Only the most precious objects were fit for the tomb of a pharaoh, and so it is likely that the metal held great spiritual significance. As is the case with the spheres of Tepe Sialk, we may never know for sure if its creator saw the meteorite falling from the sky or simply happened upon it atop the sands of the north-east African desert, but there are hints that they had at least some idea of its celestial origin.

Around the same time that Tutankhamun ruled over Egypt, a new hieroglyphic phrase came into use.

It translates into 'iron from the sky'. The phrase was used to describe all types of iron, whether celestial or strictly Earthbound, and so, while a little ambiguous, it does suggest some recognition of where this iron came from. Some ancient cultures made a connection between meteors, fireballs and the strange objects that sometimes accompany them. They probably knew that lumps of iron occasionally fell from the sky, and they held these celestial stones in reverence. They knew these objects were important. This realisation, along with many others, was severed by the Dark Ages, however, and lay forgotten for over 2,000 years. However, the idea that stones could fall from the sky was rediscovered by modern science and Europe during the Enlightenment of the eighteenth century, in a series of chance events and serendipity.

The tranquillity of a pleasant evening in the summer of 1751 in north Croatia was disturbed by a rare cosmic event. A brilliant flash of light, easily visible above the soft evening skyglow, illuminated the sky over the village of Hraščina. Almost immediately after the bright flash, a booming sound rippled across the surrounding farmland and was heard over an area of almost 2,000 square kilometres; the roaring echoes of the explosion sounded like the deep rumbling of many horse-drawn carriages. Seven eyewitnesses out on evening strolls recalled seeing two balls of fire, linked by a glowing fiery chain, fall from the sky. Some even reported two large pieces of falling rock plunging into a freshly ploughed field, and large crevasses opening where they had struck the earth. The fallen pieces were later recovered from the soft soil, one of them from a depth of almost one and a half metres. A strange black surface crusted them as if they had been charred by intense fire, and concealed the metallic nature of the objects. The long trails of smoke left behind by the falling fires were left hanging in the evening air for several hours before dissipating into the night. The event was chronicled by a local priest: 'In their ignorance, the common folk thought that the heavens had opened.'

And it is no wonder. Explosions in the sky were not an everyday occurrence in the mid-eighteenth century. The prevailing wisdom held that solid objects do not fall out of nowhere. That would be ridiculous: the heavens are perfect and without flaw, after all. Isaac Newton, one of the greatest and most influential scientists of the previous millennium, posited in his 1704 publication *Opticks* that outer space must necessarily be void of all small objects – including stones and pieces of metal – if his laws of gravity were to work. The common view held that meteors were a purely atmospheric phenomenon and had nothing whatsoever to do with the Heavens.

If there are no rocks in outer space other than the planets, the moons, and the occasional comet, then rocks could not possibly fall to the ground. Even so, seven eyewitnesses in northern Croatia gave sworn testimonies as to what they had seen: rocks falling from the sky, seemingly from nowhere.

With mind and eyes wide open

Ernst Florens Friedrich Chladni was born in 1756 in eastern Germany, and had been fascinated by physics and the natural sciences since boyhood. Because his father disapproved, Chladni wound up studying law and philosophy; he achieved his PhD in law at the age of twenty-six. Once his father died, Chladni quickly returned to his early passions. He went on to publish a seminal book on the physics of sound, *Entdeckungen über die Theorie des Klanges* (*Discoveries in the Theory of Sound*), in 1787: history now remembers him as the Father of Acoustics. Chladni is less well-known for his significant contribution towards another brand-new field of science: cosmochemistry.

Chladni was inspired by a conversation with Georg Christoph Lichtenberg, a prominent natural philosopher who had witnessed a spectacular fireball over Göttingen (north Germany) in 1791. Chladni asked what he made of the growing list of reported fireballs and the occasional reports of strange stones and metals falling from the sky. Lichtenberg responded that in his view, fireballs were not atmospheric but cosmic phenomena, originating in outer space. He speculated that the testimonies of stones and pieces of iron falling from the sky might really be genuine, though he didn't *truly* believe this to be the case himself.

This conversation set Chladni's imagination alight. He spent the following few weeks in Göttingen compiling a list of twenty-four well-documented fireballs, each observed between 1676 and 1783. Eighteen were accompanied by pieces of rock that allegedly fell from the sky, though few scholars trusted the authenticity of these claims. The rocks seemed to differ in their natures: some were stony, some were metallic, and some were a mixture. Chladni compiled their speeds, apparent sizes, flight paths, and other small details such as the number and violence of visible explosions, and the thunderous sounds made as they fell. The descriptions were astonishingly similar. Even though these accounts spanned more than a century and occurred across multiple continents, they bore a striking resemblance to one another. Adept at teasing out truths from witness accounts thanks to his legal training, Chladni deemed the accounts genuine. They were too similar not to be. What reason did the witnesses have to lie? If they were not telling the truth, how could their accounts be so similar?

Chladni shared his thoughts with the world in his 1794 publication *Über den Ursprung der von Pallas gefundenen und anderer ihr änlicher Eisenmassen und über einige damit in Verbindung stehende Naturerscheinungen*, which in English is shortened to *Ironmasses*. He argued that stones – both rocky and metallic – really do fall from the sky and are every bit as real as the Earth on which we stand. He argued that all fireballs, and the smaller lights sighted streaking through the sky called *meteors*, are caused by solid objects falling through the atmosphere at extraordinarily high speeds.

This was the first time that somebody had explicitly linked fire-balls and meteors ('shooting stars') with solid objects. It went against all conventional wisdom of the time. But Chladni went further still, concluding that the astonishing speed at which fire-balls and meteors streak across the sky rules out an atmospheric origin: the meteor stones, or meteorites, must originate from the cosmic sphere, far above the atmosphere, to be travelling at such blistering speeds. They are not of this planet. Chladni also reasoned that other strange rocks that bore a resemblance to witnessed meteorite falls but were not found in association with a fireball – those with charred and blackened surfaces – also had a cosmic origin.

This radical departure from the prevailing worldview was not warmly embraced by the scientific community. Even Lichtenberg had great trouble believing it at first. But then another stone was witnessed falling from the sky the following year, and it happened to fall in just the right place, at just the right time, on just the right person's land.

An extraordinary stone

The green rolling farmlands that stretch across the Wolds of the East Riding of Yorkshire are punctuated only by the occasional picturesque village, but one day in December 1795, the quiet was shattered by explosions emanating from the sky. The noise was heard in coastal villages fifteen kilometres away. And with the thunder still echoing across the fields, three ploughmen watched open-mouthed as a large rock fell from above and struck the ground with a dull thud. John Shipley, one of the ploughmen, was a mere eight metres away from where the rock landed. Sods of muck were hurled high into the air as it struck the soil. The twenty-five-kilogram stone, the size of a loaf, was travelling so fast that it sliced downwards through half a metre of earth and embedded itself into the solid rock below.

The owner of nearby Wold Cottage was a man by the name of Edward Topham, who was away on business at the time the meteorite fell. A playwright and founder of a scandalous newspaper called *The World*, Topham cut a flamboyant figure in late Georgian London, with his mutton-chop whiskers, unconventional sense of fashion, and charisma earning him a reputation as an eccentric (he was frequently the subject of caricaturists). But he was widely regarded as a fair and just man. He had retired with his three daughters (deemed 'the best horsewomen in Yorkshire') to Wold Cottage some years before. It was rumoured that he intended to spend the rest of his days farming, breeding greyhounds, and writing the history of his own life. His kennels were already considered the best in England, and his greyhound Snowball was praised as 'one of the best and fleetest greyhounds that ever ran'. But the meteorite interrupted all his plans and his memoir was never to be completed.

Topham returned home to find that the fallen rock which the ploughmen had carried back to his cottage had caused something of an ongoing commotion: thirty to forty people passed by each day for almost three weeks to see it, and he amassed a pile of letters requesting more details. He recorded statements from each of the ploughmen and published them, along with his own thoughts on the strange event, in the *Gentleman's Magazine*.[2] Crucially, Topham believed the ploughmen, and trusting Topham's judgement, other people believed the ploughmen, too. But the mystery of exactly how the rock had come to fall out of the sky remained; Chladni's outlandish idea that such objects originated in outer space had not yet caught on.

From Yorkshire to London

The Wold Cottage meteorite went on to become a national curiosity. In the *Yorkshire* volume of *The Beauties of England and Wales*, published between 1801 and 1815, the Wold Cottage event featured prominently. Being well connected in the capital, Topham had the meteorite sent to London to be placed on public display in the heart of the city. The exhibit was widely advertised in newspapers, including *The Times*.

For the modest price of one shilling (nearly £4 in today's money), visitors could see the strange object for themselves. They also received a copy of the testimonies of the three plough-men and a small illustration of the meteorite. When the President of the Royal Society, Sir Joseph Banks, paid his shilling to see the famous stone for himself, he noticed that it looked uncannily similar to a stone that apparently fell from the sky during a fire-ball event in Italy over a year previously. The stones were almost identical, yet they fell in different countries more than eight-een months apart.

Banks, however, was firmly of the mind that these stones had been generated in the atmosphere *by* meteors, rather than being the *cause* of meteors. His interest piqued, he tasked a talented young British chemist named Edward Howard with analysing the stones.

As a Fellow of the Royal Society, Howard was renowned for his dedicated research into synthesising new explosives for use in fire-arms (and suffered many injuries during the course of his experi-ments). Howard was loaned pieces of a further six meteorites, taking his total number of samples to eight. The stones differed in their geological characters: four were meteorites of stone, two were of pure metal, and two were a mixture of both stone and metal.

One of the iron meteorites was loaned by the Natural History Museum in London: it came from Campo del Cielo. The rock seen falling out of the sky ~ 4,000 years ago in South America was now being used for research purposes in a chemistry laboratory in Victorian London, marking a point in time where the spiritual and scientific significance of meteorites became irreversibly entwined.

Howard published his findings in 1802, and his paper remains one of the most important in the history of meteoritics.[3] It was the first systematic documentation of the chemical and geological make-up of meteorites. A few chemists had attempted to discern the chemical nature of these stones previously, but Howard's efforts were far more sophisticated. He took particular care with the stony meteorites, realising they were composed of innumerable individual grains, which he painstakingly divided into four separate components: strange rounded globules, yellow pyrites, small metallic blebs, and all sandwiched and held together by a fine 'earthy' substance made of crumbly rock. This must have been tedious work: the grains in stony meteorites are tiny. It would be like manually separating all of the poppy seeds from a bag of mixed bird food.

Howard discovered that the small metallic blebs in stony meteorites contained nickel. Nickel had been found in abundance in iron meteorites by French chemists; Howard replicated and confirmed these findings with his own set of irons and stony irons, too. It is unusual to find high levels of nickel in rocks of an Earthly origin, and so, for the first time, Howard chemically linked the stony meteorites with the iron and stony-iron meteorites. These rocks were unlike any described before them.

Their unworldly nickel-rich chemical compositions, and the distances and lengths of time between the apparent falls of similarly strange stones, pointed towards the previously laughable: a celestial origin. Howard had not only provided the first physical evidence in favour of Chladni's hypothesis, but also fashioned the field of cosmochemistry, the chemical study of material from outer space.

It still seemed unbelievable, but as Howard himself beautifully put it: 'To disbelieve on the mere grounds of incomprehensibility, would be to dispute most of the works of nature.'

Slowly, and with initial reluctance, the scientific world came round to the idea that meteors and fireballs are indeed caused by rocks falling to the Earth from space, and sometimes these rocks make it to the surface. It was fortuitous that the Wold Cottage stone happened to fall on Topham's land when it did. Had it fallen on somebody else's land – perhaps somebody with less of a gift for publicity – it may have wound up being used as a doorstop. (This really did happen to a meteorite named Lake House: it unceremoniously sat on the doorstep of an Elizabethan country house, from which the meteorite took its name, in Wiltshire (south-west England) for almost a century before its celestial origin was confirmed by scientists at the Natural History Museum in London.)

Today, the exact spot where the Wold Cottage meteorite made landfall is marked by a tall monument, commissioned by Topham himself and built from reddish-brown brick. On it sits an ornamental stone tablet, into which the following inscription is chiselled:

Here
On this spot, *Dec 13th, 1795,*
fell from the Atmosphere
AN EXTRAORDINARY STONE
In Breadth *28* Inches,
In Length *30* Inches,
and
Whose Weight was *56* pounds
THIS COLUMN
In Memory of it
was erected by
EDWARD TOPHAM
1799.

By the middle of the nineteenth century, all but the stubbornest intellectuals had accepted Chladni's hypothesis. But there remained a huge problem: where exactly in outer space do meteorites come from in the first place?

Many other places

Chladni hypothesised that meteorites originated not merely outside the Earth's atmosphere, but outside our Solar System entirely. He ascribed an interstellar ('between the stars') origin based on the lightning speed at which they enter our atmosphere. Chladni's other theory was that meteorites may originate from a destroyed planet, but no evidence of large planetary fragments had been seen by telescopic observations of the night sky. It wasn't long, however, before another hypothesis on the origin of meteorites was proposed.

In 1802, the same year that Howard published his work on the chemical nature of meteorites, Pierre-Simon Laplace, a French mathematician and astronomer, popularised the hypothesis that these meteorites originated closer to home. He suggested that they came from the Moon. Eruptions of Lunar volcanoes had been 'observed' by the German-British astronomer William Herschel in 1787 (this apparent observation, in time, turned out to be an error). Laplace hypothesised that if the powerful volcanic forces seen on the Earth are also active on the Moon, then ejecta thrown out by these volcanoes could be projected into space and onto the Earth. It seemed watertight. The hypothesis was so popular that in the Yorkshire volume of *The Beauties of England and Wales* guidebook, the Wold Cottage meteorite was described as a piece of the Moon.

In the meantime, the inventory of known meteorites was growing. By the middle of the nineteenth century over 150 celestial stones were housed in museum collections and cabinets of curiosities of the wealthy. It was around this time that the Lunar origin hypothesis took a fatal hit. Benjamin Apthorp Gould, an American astronomer, in 1859 published his calculations on the probability of a piece of ejecta from a Lunar volcano reaching the Earth: the chances were less than one in a million. Gould's calculation showed that for every piece of Lunar lava that landed on the Earth, over one and a half million were also ejected into deep space. If the 150 or so meteorites that had fallen in the last few centuries really did originate from the Lunar surface, the Moon should have visibly shrunk in that time due to the massive amount of material ejected and lost by volcanoes. The Moon was not shrinking, however. It turned out that the answer to the conundrum of where meteorites originate was hiding within another quandary facing astronomers at the time: there was a 'missing planet'.

A useful yardstick in astronomy is the 'astronomical unit', or the 'au'. It is roughly equal to the distance between the Sun and the Earth, about 150 million kilometres. An astronomical unit is a long way. Light, the fastest moving thing in the Universe, takes eight minutes and nineteen seconds to traverse 1 au: for comparison, it would take you just over 150 years to drive this distance in a car. Mercury, the planet closest to the Sun, orbits the Sun at a distance of 0.4 au. The next planet out, Venus, orbits 0.7 au from the Sun. Earth orbits at 1 au. The Red Planet, Mars, orbits at just over 1.5 au, and then nothing but void, until Jupiter, which orbits the Sun at 5.2 au. The gap between Mars and Jupiter had bothered astronomers for centuries. Many believed that an undiscovered planet lay hidden there.

On New Year's Day 1801, Italian astronomer Giuseppe Piazzi was compiling a catalogue of the position of stars at his telescope in Sicily when he noticed something strange in the sky. It was a point of light, but it had a strange colour and didn't quite look like a star. (At this point, Piazzi had been working on his stellar catalogue for nine years.) Puzzled, he searched for this not-quite-right-looking star the next evening only to find that its position had shifted slightly. This was odd. Stars do not move from one night to the next.* He looked for the point of light on the third night. It had moved again. Piazzi realised that there was no way that this object could be a star. This is a fine example of how some of the best scientific discoveries come from 'hmm, that doesn't look quite right' moments.

Piazzi initially thought that this new object in the sky was a comet. It was far too small to be a planet, appearing only as a tiny point of light even with the most powerful magnifying telescopic lenses. But subsequent observations by himself and fellow astronomers failed to detect the characteristic cloud, the fuzziness, that tends to envelop comets. Its orbit around the Sun, too, was not at all comet-like. Comets have highly elliptical orbits: they orbit the Sun along paths that resemble elongated or 'squashed' circles, whereas the orbit of this new object was almost circular. It was a planet-like orbit. What is more, it was orbiting the Sun in the void between Mars and Jupiter. Piazzi had accidentally discovered the 'missing planet'. Following the long tradition of naming celestial objects after deities – a practice that highlights the spiritual nature of the night sky – he named it Ceres after the Roman goddess of agriculture.

Only one year later, Heinrich Wilhelm Matthias Olbers, a German astronomer, spotted an object in the sky with similar properties. It moved across the sky from night to night, and its orbit was too circular for it to be a comet. The cometary cloud-like haze was missing, too, and it orbited the Sun in the

---

* This is not *strictly* true. All of the stars in the sky are perpetually moving relative to each other but at a rate almost imperceptible on a night-by-night basis.

same part of the Solar System as Ceres, right between the orbits of Mars and Jupiter. He named the new planet Pallas after the Greek goddess of wisdom. Pallas, like Ceres, was minute, appearing as little more than a speck of light against the black backdrop of space. It was strange, however, that Ceres and Pallas orbited the Sun at roughly the same distance. Astronomers had predicted just one 'missing planet'; nobody had predicted there would be two. All of the other known planets were the dominant object in their orbital lane, but Ceres and Pallas seemed to break this rule. Olbers hypothesised that they were the surviving fragments of a planet that was shattered into pieces, perhaps by a catastrophic cometary impact or an internal explosion. He predicted that more fragments would soon be found.

Herschel summarised the discoveries and characteristics of the two new 'planets' in a Royal Society publication.[4] They looked like stars through a telescope, they were small like comets but were not shrouded by the cometary haze, and they orbited the Sun like planets. Sharing similarities with stars, comets, and planets, but being distinct from each, Herschel suggested that they could be a new class of astronomical object. Thus, he coined a new word by which to call them: 'asteroids', from the Greek words ἀστήρ- (astir) and -εἶδος (eidos) which together mean 'star-like'. The word did not immediately catch on, however, and they were still widely referred to as 'planets' or 'planet fragments' by astronomers.

A third asteroid, named Juno, was discovered in 1805 by German astronomer Karl Ludwig Harding. Olbers discovered the fourth asteroid (his second), Vesta, in 1807. Four small 'planets' in the gap between Mars and Jupiter further bolstered the fragmented planet hypothesis proposed by Olbers. Clearly there was something strange going on between the orbits of Mars and Jupiter.

Chladni was elated. In *Ironmasses* he had, in what was pure speculation, suggested that meteorites might be the small fragments of a destroyed planet. Asteroids were physical proof that his hypothesis might bear some truth. Perhaps meteorites were shrapnel from a planetary cataclysm that fell upon the Earth. What is more, several astronomers had reported variations in the brightness of asteroids: they ascribed this to their presumably fragmented and irregular shapes. If they were indeed pieces of a broken planet, it made sense that they would resemble shards and reflect different amounts of light as they tumbled and rotated in their orbits.

No new asteroids were spotted for almost forty years. But between 1845 and 1855 there was a burst of asteroid discovery when a further *thirty-three* asteroids were catalogued, taking the total to thirty-seven. Ten years later the total was at eighty-five. By now most people recognised that they could not *all* be planets, and the term 'asteroid' had seeped into common usage. The space between the orbits of Mars and Jupiter started being called 'the Asteroid Belt': the asteroids appeared to exist as a wide belt of rocky debris orbiting the Sun. The belt spanned from around 2 au to around 4 au, a distance of almost 300 million kilometres, which is twice the distance of the Sun to the Earth. The orbit of the 'missing planet' was shaping up to be a vast interplanetary field of asteroids.

Remarkable chasms

As more and more asteroids were discovered and their orbits calculated, patterns began to emerge within the belt. American astronomer Daniel Kirkwood noticed concentric gaps in the belt – since termed 'Kirkwood gaps' in his honour – that were void of asteroids. He described these gaps as 'remarkable chasms' in 1866. The Asteroid Belt wasn't simply a wide ring of chaotic debris orbiting the Sun: it was ordered into concentric rings. Kirkwood correctly attributed the gaps in the rings to gravitational interactions with the biggest planet in the Solar System, Jupiter. As the asteroids dance around the Sun along with the planets, certain regions of the belt are in an 'orbital resonance' with Jupiter. A consequence of Newton's Law of Universal Gravitation is that the distance a planet, asteroid, or comet lies from the Sun will dictate the speed at which it orbits. The further away from the Sun something orbits, the slower it goes. Orbital resonances in the Asteroid Belt happen when the orbital period of an asteroid and the orbital period of Jupiter can be expressed as whole-number ratios.

Imagine the Solar System resembles a clock face, with the Sun at its centre and the planets and asteroids orbiting at different distances from it. On our clock, Jupiter orbits along the outer edge of the clock face. Now imagine that there is an asteroid closer to the centre of the clock face (closer to the Sun): this asteroid will complete one orbit faster than Jupiter, which lies further out. As we watch, we realise that this particular asteroid completes one trip around the clock face (one orbit around the Sun) twice as fast as Jupiter. One orbit for Jupiter; two orbits for the asteroid. This is called a 2:1 orbital resonance. Every second asteroid orbit, both Jupiter and the asteroid will be at twelve o'clock on the clock face. At this position, the strong gravitational field of Jupiter tugs slightly on the asteroid, causing its orbit to become ever so slightly elliptical. After hundreds of thousands of orbits, the small gravitational tugs at twelve o'clock add up and the resonance heaves the asteroid into a chaotic orbit. Similar resonances (and therefore chasms) are found at 3:1, 5:2, 7:2, and 7:3.

Their chaotic orbits may move an asteroid into the safety of a more gravitationally stable part of the belt. Some asteroids may be ejected from the belt entirely, sent careening either inwards towards the Sun or outwards into the frigid outer regions. Others will be sent on a cataclysmic collision course with other aster-oids which will produce a deluge of shrapnel. Whatever their fate, any asteroids found within a region of orbital resonance will be swiftly budged. Thus, specific lanes in the belt become emptied of asteroids, through little more than an elegant conspir-acy of orbits.

Orbital resonances emptying parts of the belt provide the means by which asteroids – and small fragments of asteroids – can collide and escape into other regions of the Solar System. If the orbit of an asteroid or asteroid-fragment is perturbed such that it becomes Earth crossing, there is the potential that it will be swept up by the Earth as it journeys round the Sun. The tantalising possibility that meteorites may represent such fragments from the Asteroid Belt was not lost on scientists of the time.

Asteroid shrapnel

While the astronomers were looking upwards through their telescopes, geologists were looking downwards through their microscopes. In the middle of the nineteenth century, French geologist Adolphe Boisse thought he had found evidence that meteorites originate from a fragmented planet, in line with the asteroid–origin hypothesis. He arranged them in a sequence of decreasing density so that they resembled the interior of a planet such as the Earth: iron meteorites in the centre, representing the metallic core, overlain with the hybrid-like stony irons, and then the stony meteorites encasing the lot, representing the outermost rocky mantle and crust. That meteorites resemble the layers of a large planet was taken as compelling physical evidence that asteroids were definitely the pieces of a fragmented planet, and that meteorites originate from them.

A huge problem still remained. Chladni had pointed out in *Ironmasses* that the speed at which fireballs and meteors streak across the sky precluded a Solar System origin: he (and eventually others) thought they must originate from interstellar space to be travelling so fast. This was at odds with the asteroid-origin hypothesis. While the scientific community had universally come round to the idea that fireballs and meteorites are caused by rocks falling through the Earth's atmosphere from space, and that some of them survive their fiery descent to the Earth's surface, it wouldn't be until the middle of the twentieth century that the problem of exactly where they originate was solved.

Throughout the 1930s and 1940s, there was a focused effort to capture the atmospheric entry of fireballs on camera to better understand their flight paths through space. If their trajectories were precisely recorded, their speeds and orbital paths could be calculated, and so the question of whether meteorites were interstellar objects or originated from within our Solar System would be solved. Capturing a fireball on camera is a case of good fortune and involves plenty of waiting around. But patience paid off. Scientists at astronomical observatories in the United States eventually succeeded in capturing fireballs passing overhead on long-exposure photographic film. Their calculations, based on the speed and direction of atmospheric entry, showed that fireballs were produced by stones that were in orbit around the Sun. They do not originate in inter*stellar* space: they originate in inter*planetary* space. They are of this Solar System.

By the mid-1950s, motivated by Cold War tension, Czechoslovakia, like many other nations, had established a network of skyward-pointing cameras to capture the flight paths of artificial satellites passing overhead. Monitoring space was a matter of national security. In April 1959, at a monitoring station in

Příbram, a small town fifty-five kilometres south west of Prague, several cameras simultaneously captured the overhead passage of a fireball. The fireball was witnessed by people on the ground, too, over an area of almost eight thousand square kilometres, and it illuminated the night sky as it fell. This marked the first time in history that a fireball had been caught on film by more than one camera. Because the flight path of the fireball was captured from more than one angle, the forwards trajectory of the fireball could be triangulated and calculated with high precision. This allowed a prediction of its landing site to be made. The prediction proved sound, and within weeks a freshly fallen stony meteorite the size of a large apple, complete with charred black exterior, was found. Three more pieces of the same meteorite were found in the following months, indicating that the stone had broken into pieces as it plunged through the atmosphere.

Not only was the forward trajectory calculated, but the backward trajectory was calculated too. Based on the incoming speed and direction of the fireball, the stone's path through space was determined: it arrived on Earth on a highly elliptical, Earth-crossing orbit. The Příbram meteorite originated from the outer region of the Asteroid Belt. This was the first hard evidence that celestial stones really do originate from the belt of asteroids between the orbits of Mars and Jupiter.

After decades of watching the skies with a network of camera systems across the globe and re-tracing the paths taken by fireballs through space, many more meteorites have had their orbits accurately back-calculated based on the speed and direction in which they entered Earth's atmosphere, including Innisfree that fell in Alberta (Canada) in 1977, Morávka in the Czech Republic in 2000, and Park Forest in Illinois (USA) in 2003. Each, like Příbram, was caught on camera and pieces of the falling stones were collected. In the case of Morávka, an apple-sized piece of the meteorite struck a spruce tree before landing in (an exceptionally lucky) somebody's back garden.

There are plenty of asteroids from which meteorites can come, too. Since Piazzi accidentally discovered Ceres in 1801, hundreds of thousands of asteroids have been catalogued within the Asteroid Belt. Ceres, as well as being the first asteroid to be discovered, is by far the largest: at just shy of 1,000 kilometres across, it is as wide as Great Britain is long. Pallas and Vesta are both about 500 kilometres across, about the size of England. It is estimated that there exist somewhere between one and two million asteroids larger than one kilometre across, and we have a long way to go before we map them all. Large asteroids are the exception rather than the rule, though, and are exceedingly rare in comparison to the smaller ones: there are probably billions upon billions of asteroids smaller than one kilometre across, and they range in size from one metre across downwards.

Silently circling the Sun, the asteroids send forth fragments of themselves into the inner Solar System when they collide and eject fragments at great speed, or stray into an orbit too close to one of Kirkwood's great chasms of gravitational resonance. Through interplanetary space these fragments travel. Most that stray into the Earth's path are minute, dust- to pea-sized, and so are completely obliterated as they fall through the atmosphere. For the larger pieces, though, their survival through a blazing atmospheric entry and landing on Earth marks a new chapter in their long history. Most meteorites lie undiscovered and, through the passage of geological time, become incorporated into the Earth's surface and are turned into Earth stuff. Only a tiny fraction of them is recovered by Earth's curious inhabitants.

From the meteorites, we have learnt about the asteroids from which they originate in intimate detail, and in doing so have turned them from faint points of star-like lights in the sky into worlds. They are places with their own histories and their own stories to tell. Within them is written the earliest chapter of the Solar System's history, the instructions of how to assemble a planetary system, and the ingredients needed to build new worlds.

Much of the remainder of this book is concerned with the tales from antiquity bound within these stones, but before we delve into the first pages of our Solar System's story, we must find ourselves a meteorite.

2

EARTHFALL

For somebody to witness a meteorite falling to the Earth's surface and spot exactly where it lands is exceedingly rare. Meteorites that are seen falling and are recovered shortly after they land are called 'falls', and there are less than 1,200 of them out of the approximately 60,000 known meteorites.[1] That is less than one in fifty. Falls are highly prized, and are usually the most important meteorites in several aspects: they are culturally important because with their fall comes an excited flurry of eyewitness accounts and fabulous stories, and they are scientifically important because they best preserve the geological tales scribed within.

Until they land on the Earth's surface, meteorites exist in the inert vacuum of space, and the absence of any gases to react with preserves them, immutable, for billions of years. Even during their violent atmospheric entry, they remain largely pristine: only the outer skin of the meteorite is superheated, and it is quickly stripped away before the heat has time to work its way inwards. As soon as a meteorite lands on Earth, however, it immediately begins to be weathered: it is attacked from all sides by the oxygen in our atmosphere, rain, and the legions of microbial life that coat the surface of the planet. Geological processes, while working to record new stories within Earth rocks, act to overwrite the stories written within meteorites. Being unspoiled, falls are often worth many times their weight in gold.

Meteorites fall roughly evenly across the Earth's surface, albeit with a slight increase at equatorial latitudes, and so there is about as much chance of one falling in a field in Scotland as there is in a similar sized field in Australia or Peru. Earth is mostly covered in ocean, and so most meteorites sink into a watery abyss and are forever lost. The meteorites that are lucky enough to touch down on dry land take their name from the place they fell. They therefore inherit names that encompass the full spectrum of graphical place names, ranging from the quaint (Wold Cottage), to the tongue-twisting (Millbillillie), to the amusing (Camel Donga).

Some 40,000 tonnes of extraterrestrial material falls to the Earth's surface each year. So, why isn't the surface of the planet covered in a thick layer of meteorites? It is a matter of size. The Earth is enormous: it is the biggest rocky planet in the Solar System, and with a surface area of half a billion square kilometres, even 40,000 tonnes of material is spread pretty thinly, like too little butter over too much toast. The equivalent of one single teaspoon of extraterrestrial rock is sprinkled over an area the size of St James's Park in central London each year. It is barely noticeable.

Not all of the extraterrestrial material falling to the Earth is stone-sized. Most of it rains down as microscopic particles of rock called 'cosmic dust', only visible with the aid of keen eyes and a microscope. Only a tiny fraction falls as anything that could be considered 'stone-sized', and meteorites the size of the falls of Wold Cottage or Hraščina are exceptionally rare. Mammoth meteorite falls that are car-sized, like the ancient fall of Campo del Cielo, are once-in-a-human-lifetime events, and even then they mostly plunge into the ocean and are never recovered.

The almost 59,000 meteorites that were not witnessed falling are called 'finds'. These meteorites lie on the Earth's surface until they are discovered and moved to the safety of a curation facility, where they are sometimes stored in dehumidified air and steady temperatures to prevent them from reacting with moisture in the atmosphere. Some facilities preserve ultra-precious meteorites under high-vacuum or a pure nitrogen atmosphere to prevent them from rusting.

Finds may linger on the ground for tens or hundreds of thousands of years before they are discovered: if they are not discovered in time, they are weathered away beyond recognition and become incorporated into the sands and soils of Earth, reduced to nothing more than gravel. It is unsettling to think that rocks that survived in interplanetary space from the formation of the Solar System billions of years ago can be destroyed by spending less than one million years on Earth's surface.

Like falls, finds are scientifically valuable, though care must be taken when analysing them because they are often riddled with defects: metals oxidise to rust, new minerals are formed in complex water–rock interactions by percolating rainwaters, and veins of salt precipitate as fluids infiltrate cracks. Teasing apart features that are indigenous to the meteorite, and those that are Earthly, is sometimes a huge challenge.

We have gleaned a wealth of information from studying meteorites, in part thanks to the sheer volume we have available to study. But it wasn't always this way: Howard only had eight meteorites when he performed his important analyses early in the nineteenth century, and even by the mid-twentieth century there were less than 2,000 in total. All that was changed, however, by a chance discovery in one of the remotest corners of our planet that sparked a meteorite gold rush.

Cold desert finds

In the southern hemisphere's summer of 1969, a team of Japanese scientists were embedding monitoring stations into the East Antarctic Ice Sheet to track the flow and creep of the glacial ice.

On the afternoon of 21 December, several of the team produced strange-looking stones they had found sitting atop the ice. The stones were each crusted with a black outer skin. Masaru Yoshida, a geologist out on the ice, suspected these stones might be meteorites and encouraged the rest of the team to keep an eye out for more. A further six were found over the following ten days, taking the total to nine. These peculiar dark rocks stood out on the pale blue glacial ice, and so were easy to spot. The samples were bundled in packing tape, wrapped in a cloth, and returned to Japan in a tin container for analysis.

Back in Japan, Professor Masao Gorai, the geologist tasked with characterising the samples, sliced the stones open to take a close look under a microscope. It didn't take long for him to recognise all nine as meteorites. Their textures and geological make-ups differed wildly: while each was stony in nature (none contained significant amounts of metal), at least five distinct types of meteorites had been found. They must have each fallen separately, one by one, probably over tens of thousands of years, rather than together as a shower of fragments of a single stone. To find so many individual meteorite falls in one place seemed implausibly serendipitous. Impossible, even.

Then, in 1973, quite by chance, Japanese scientists found a further dozen meteorites on the same patch of ice. These meteorites were also all stony, yet at least five different types (three not seen in the 1969 samples) were present, each containing a different tapestry of minerals. Excited at the prospect of a meteorite bonanza, in 1974 a dedicated party was sent from Japan back to the East Antarctic Ice Sheet to hunt for more.

They discovered more than 660 meteorites in just one fortnight, increasing the total number of known meteorites by about one-third. It was unprecedented. Over two dozen different types of meteorites were found, too, most being varying types of stony meteorites, but amongst them was one stony-iron meteorite.

This, it turned out, was just the tip of the iceberg.

Japanese scientists immediately recognised that there was something unusual at work bringing meteorites together, because there was no way they could have all fallen separately, one by one, onto the same small patch of ice sheet. It seemed impossible. How could it be?

The answer lies within the ice. Over time, meteorites fall at random onto the East Antarctic Ice Sheet – the largest ice sheet on the planet – as they do over the entire planet. When they fall they become buried by fresh snowfall, deeper and deeper, until they are fully incorporated into the ice sheet and become entombed within it. As millennia pass by and the crushing mass of overlying snow thickens, the air is squeezed out of the ice by the immense overlying weight, turning it from brilliant white into shocking pale blue. The meteorites become encased in this blue prison. As the massive ice sheet flows towards the edges of the continent, the meteorites held within are taken along for the journey as if on a natural conveyor belt.

Occasionally, the flowing ice is hindered by submerged chains of mountains beneath the ice sheet. When the ice sheet buckles up against these barriers, it is forced upwards: old ice from deeper down is squeezed to the surface. Instead of flowing onwards towards the ocean at the edge of the continent and carrying the encased meteorites to a watery end, the standing ice is blasted and stripped away by high-speed polar winds. The heavy meteorites are mostly left behind. The powerful winds prevent fresh snow from accumulating, too, and as the surface ice is blown away, more is pushed upwards from below to take its place. Over millennia, as masses of ice are stripped away, the meteorites pile up in phenomenal numbers.

Within a year of the Japanese expedition, word of a vast hoard of meteorites in Antarctica had circulated throughout the scientific community, and the United States, still excited from its recent *Apollo*-era heyday, decided to hunt for them too. A meteorite Gold Rush ensued.

By 1980, joint Japanese-United States annual expeditions had uncovered almost 5,000 new meteorites, doubling the total number of known meteorites worldwide. At the time of writing, almost 40,000 meteorites have been collected from the East Antarctic Ice Sheet, and hundreds more are discovered each year. To this day, annual expeditions continue to scour the blue ice of the Antarctic ice sheet in search of meteorites.

The treasures collected in Antarctica set meteorite research alight. No longer confined to studying ultra-precious historic falls housed in the world's great museums, cosmochemists today have access to a vast new Antarctic collection containing many different types of meteorites.

Hot desert finds

While two-thirds of all meteorites come from the world's largest cold desert, a further fifth come from the world's largest hot desert: the Sahara.

The Saharan aridity preserves fallen meteorites for hundreds of thousands of years, allowing them to accumulate on the sand in large numbers as they intermittently fall from above. Their blackened exteriors jump out against the bright sand, and in an environment otherwise void of rocks and boulders (and trees and bushes), there is usually only one place from where a stone could have come: the sky.

Knowing that scientists and amateur collectors will often pay handsome sums of money for these strange stones,[2] the peoples of the Sahara collect and sell them at the markets that line the edges of the desert, and in the case of rare meteorite types they make vast sums of money.

But there is a dark side to Saharan meteorites. Many countries in the northern region of Africa have tight controls and regulations overseeing the trade in these natural wonders, and thus many meteorites leave the continent illegally. Having changed hands numerous times, cosmic contraband makes its way into private collections, curation facilities, and scientific institutions across the world, often without paperwork or any traceability. These ethical considerations deter some institutions from housing meteorites found in Africa – the Natural History Museum in London being one such place – but spurred on by scientific enquiry, many institutions plough onwards and conduct vast amounts of research on Saharan meteorites. For better or worse, the trove of meteorites collected in the Sahara Desert has greatly advanced our understanding of the asteroids from which they originate. It is a double-edged sword.

Together, deserts hot and cold are Nature's way of collecting, storing, and preserving meteorites and the stories written within them. The wondrous ability of the East Antarctic Ice Sheet to gather and draw together vast numbers of meteorites, as if drawing together the scattered pages of a book that fell from the sky, makes it an Earthly gateway into the rest of the Solar System and into the deep past. Stories are stowed away in the ice or on the hot sand, like old books patiently waiting on a shelf, for up to one million years before they are discovered by explorers. Deserts are some of the great libraries of Nature.

Systematics

It was clear early on to Howard and the other pioneers of cosmo-chemistry that meteorites come in three different varieties – stony, stony iron, and iron – but being limited to the few dozen specimens housed in the museums across Europe, they had no way of knowing about the true variety of meteorite types.

From the vast collection of meteorites amassed over the

previous two centuries, coupled with the advanced tools of science, the enormous diversity of meteorites has been laid bare. Thus, a picture of the asteroids from which meteorites originate has come into sharp focus. These stones are not random pieces of asteroid: there is an underlying system by which they can be organised and classified into families. Whilst almost all meteorites are encased within a charred black fusion crust, once they are sliced open using diamond-encrusted sawblades, their geological characters are revealed, and they are as varied and as wonderful as the multitude of rocks that make the Earth.

We now know that the three main varieties of meteorite can be divided further into well over forty different groups based on similarities in their geological characteristics. Stony meteorites can be subdivided into at least thirty different groups, stony irons into at least six, and irons into at least fourteen. Some stony meteorites are composed of a mosaic of igneous crystals that formed from cooling magma; others are so rich in water and complex carbon-based molecules that they have a pungent asphalt-like smell. Some stony-iron meteorites consist of penny-sized bottle-green crystals of olivine – a semi-precious gemstone colloquially known as 'peridot' – set in a skeleton of metallic iron. Some iron meteorites are comprised of tiny needles of metallic minerals; others are made of metal in the shape of long, interlocking fingers. There is a startling diversity of rock types originating from the Asteroid Belt.

Such a startling number of distinct meteorite groups immediately casts a shadow on the 'shattered planet' hypothesis: how could so many different types of meteorite all originate from the fragments of one planet? One could point towards the diversity of rocks here on the Earth: it is true, after all, that there are thousands upon thousands of different types of rock forged by the geological forces that continuously shape our home planet. Meteorites contain within them the answer to this question, but to read them we must use one of the most powerful tools in cosmochemistry: isotopes.

Atoms – the chemical building blocks of matter – each house a dense nucleus comprising two subatomic particles: protons and neutrons. The chemical nature of an atom is dictated by the number of protons, which varies from one (hydrogen, element number one) to ninety-two (uranium, element number ninety-two) across the periodic table of the elements. Protons define an element: for example, neon always has ten protons, iron always has twenty-six protons, platinum always has seventy-eight protons, and so on. But the periodic table tells only half the story, because the number of neutrons in an atomic nucleus varies, too.

Unlike protons, which have a positive electrical charge, neutrons have a charge of zero and so have a negligible effect on the chemical behaviour of an atom. A neutron has about the same mass as a proton, however, and so by varying the number of neutrons in an atomic nucleus, the mass of an atom can be changed. Atoms of the same element with varying numbers of neutrons in their nucleus are called isotopes, from the Ancient Greek ἴσος (iso) meaning 'same' and τόπος (topos) meaning 'place': isotopes occupy the 'same place' on the periodic table because they are the same element, differing only in mass.

Element number eight

Life-giving oxygen, the second most abundant gas in the Earth's atmosphere, is, perhaps surprisingly, the most abundant element in most rocks. As an essential component of the crystal structures making up a great number of rock-forming minerals, there is more oxygen locked away within the rocks beneath your feet than there is in the air we breathe. Being present in large quantities in meteorites, and possessing a unique combination of important chemical properties, it is without doubt one of the most important elements in cosmochemistry. Oxygen is a faithful and powerful storyteller.

Oxygen presents itself as three different isotopes: oxygen-16, oxygen-17, and oxygen-18, which in shorthand we write as ^{16}O, ^{17}O, and ^{18}O, respectively. Each oxygen isotope has, by definition, eight protons tucked away in its nucleus (oxygen is element number eight in the periodic table) but a varying number of neutrons: eight neutrons in ^{16}O, nine neutrons in ^{17}O, and ten neutrons in ^{18}O. All being oxygen, they are chemically identical, but being different isotopes, they have different masses: ^{18}O is the heaviest of the three, and ^{16}O is the lightest.

By far the most common isotope of oxygen is ^{16}O. If one sat and painstakingly counted out 10,000 individual oxygen atoms from the water molecules (H_2O) in a bucket of sea water, all but around twenty-four of them would be ^{16}O. Of those twenty-four, twenty would be ^{18}O, and the remaining four would be the rarest of the three isotopes, ^{17}O.

The relative proportions of the three isotopes of oxygen vary slightly from place to place on the Earth by tiny, but nonetheless measurable, amounts. For example, when water is warmed up, say in the case of a steaming-hot cup of tea, the light isotopes are evaporated more easily than heavy isotopes. In the case of water – which is an oxygen-bearing molecule, H_2O – the vapour rising from the surface of the tea as it evaporates will be slightly enriched in ^{16}O and (to a lesser extent) ^{17}O compared with the water left behind in the cup. Being lighter, the ^{16}O is evaporated more easily than ^{17}O, and ^{17}O is evaporated slightly more easily than ^{18}O.

In fact, it is half as difficult to evaporate ^{17}O compared to ^{16}O as it is to evaporate ^{18}O compared to ^{16}O, because the difference in mass between ^{17}O and ^{16}O is half the difference in mass between ^{18}O and ^{16}O.*

---

* You can check this for yourself: The difference in mass between ^{17}O and ^{16}O is 1 ($17-16 = 1$) and the difference in mass between ^{18}O and ^{16}O is 2 ($18-16 = 2$). The relative difference in mass, therefore, is $1 \div 2 = \frac{1}{2}$.

This means that the difference in the relative change in the amount of ^{17}O compared to ^{16}O in the vapour will be half as much as it is for ^{18}O compared to ^{16}O. If we were to measure the relative difference in the proportion of ^{17}O to ^{16}O, and of ^{18}O to ^{16}O, in many things (from air, to water, to rocks, to humans) from many places across the Earth, and then plot them against each other on a graph, they would fall along a straight line with a steepness of ½. We call this line the 'terrestrial fractionation line'. Everything on Earth plots on this line.

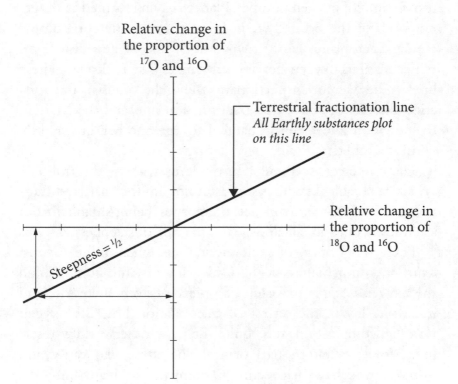

A sketch of an oxygen isotope plot commonly used in cosmochemistry. The blends of ^{16}O, ^{17}O, and ^{18}O in all Earthly substances – from ocean water, to air, to rocks – lies along the terrestrial fractionation line. Substances from asteroids, comets, and other planets do not lie along this line.

As oxygen is shuffled around the planet from place to place, whether it be water vapour rising from a cup of tea, or the melting of old rocks and crystallising of new ones, its isotope composition always changes in this systematic and orderly way. Everything gets smeared out along the same line of steepness ½; the relative change in the proportion of ^{17}O and ^{16}O changes half as much as ^{18}O and ^{16}O. Everything falls on the terrestrial fractionation line. Nature is systematic and predictable.

Different parts of the Solar System have slightly different proportions of oxygen isotopes. Planets, having formed in different parts of the Solar System, inherit these different isotopic blends. Each planet has its own unique line of steepness ½ on which all of its oxygen-bearing substances (like rocks) fall. These lines – the Mercurian fractionation line, the Venusian fractionation line, Earth's terrestrial fractionation line, and the Martian fractionation line – run in parallel to one another, but slightly vertically offset.

These lines of steepness ½ provide us with a powerful and, crucially, testable, hypothesis: if asteroids are the shattered fragments of a once-whole planet, meteorites, being shrapnel from asteroids, should all fall along a unique line of steepness ½.

There was a flurry of excitement, then, when in the 1970s, a team of cosmochemists at the University of Chicago measured the oxygen isotopic make-up of a piece of stony meteorite and it did *not* lie on the terrestrial fractionation line.[3] Its oxygen isotopic composition was, quite literally, otherworldly, testament to its extraterrestrial origin. The stone was vastly and disproportionately enriched in ^{16}O compared with anything that had ever been measured on the Earth. This measurement sparked a rush of interest in the oxygen isotopic make-up of meteorites that continues to this day and, from research conducted in the intervening decades, a clear picture of the nature of asteroids has emerged.

Rather than lying on a single line of steepness ½, meteorites group into numerous distinct clusters. This is entirely inconsistent with having originated from a once-whole planet – from which all rocks would lie on their own unique line of steepness ½ – and compelling evidence that meteorites do not originate from a single world. They originate from many distinct and unique rocky bodies, each with a unique blend of ^{16}O, ^{17}O, and ^{18}O.

Asteroids are not fragments of a shattered planet; they are fragments that never formed a planet in the first place. They were always separate; never whole. The asteroids have always been a band of solitary wanderers.

Oxygen has further tales to tell. It quickly became apparent that meteorites grouped together based on their face-value geological similarities also have identical oxygen isotope compositions. Each distinct group of meteorites comes from its own separate asteroid, and together they contain the story of their parent asteroid's geological history. Far from being a homogenous band of identical celestial debris, asteroids are as rich and diverse as meteorites themselves.

But before we discover their stories, it is worth turning our attention briefly towards the ~ 400 meteorites that originate from a more familiar planetary body.

Closer to home

Out on the East Antarctic Ice Sheet, on 18 January 1982, explorers on the annual American-led meteorite hunt braved the pinching cold. Just before returning to camp, one explorer spotted a dark, walnut-sized rock atop the blue glacial ice, the 373rd and final meteorite to be discovered that season. It was immediately clear that this was no ordinary meteorite: in places the fusion crust had fallen away to reveal sharp-cornered, centimetre-sized fragments (known to geologists as 'clasts') of a white rock surrounded by much finer fragments of dark-black rock. It was quite unlike any other meteorite known at the time. It was bundled away in the customary sterile packaging and, along with the other 372 meteorites collected that year, was returned to NASA's Johnson Space Center in Houston (USA) for classification. It was named Allan Hills 81005,[4] and, as is customary for unusual-looking Antarctic meteorites, a thin wafer was sliced off the stone and made into a thin section. Thin sections are a tool of the trade for geologists: delicate slithers of rock that are mounted onto a glass microscope slide and polished down to a thickness of thirty micrometres.* When rock is sliced this thinly it easily allows light to pass through it, and the optical properties of the light can be used by geologists to identify minerals within the rock and describe its geological character.

The thin section of Alan Hills 81005 was analysed at the Smithsonian Institution in Washington DC, and cosmochemists found that the white clasts were fragments of a mineral named anorthite and the dark clasts were fragments of rock named basalt. The clasts were bound together by a dark-brown glass, like fruit bound together and encased by the sponge in a plum pudding. While rare on Earth, anorthite is one of the most abundant minerals on the surface of a celestial body that graces the skies of Earth on a daily basis: the Moon. This sparked the tantalising possibility that this meteorite had originated a little closer to home.

It quickly became clear that the chemical, geological, and isotopic make-up (oxygen isotopes amongst them) of Allan Hills 81005 was identical to the rocks picked up from the Lunar surface and returned to the Earth by the *Apollo* astronauts a

---

* There are 10,000 micrometres in one centimetre. For comparison, a human hair is typically around 100 micrometres thick.

decade earlier. Allan Hills 81005 was a piece of the Moon. Laplace, the French mathematician and astronomer who 180 years earlier had suggested a Lunar origin for meteorites, turned out not to be entirely wrong after all.

Even with an unaided eye the main geological provinces of the Moon can easily be picked out from the Earth. Bright white and light-grey terrains envelop clusters of black-grey patches that the astronomers of old poetically named the Lunar *terrae* (Latin for land) and the Lunar *maria* (Latin for sea), respectively. While wrong about the detail, they were right in essence: the Lunar *maria* are indeed seas, but not seas of water. They are the crystallised remnants of once-molten seas of basalt, a dark grey-black igneous rock that forms as lava cools and solidifies, and a rock type that is incredibly common on Earth. The volcanic islands of Hawaii and Iceland here on Earth, and most of the ocean floor, are made of basalt. Glowing expanses of red-hot basalt filled circular impact craters on the Moon billions of years ago, which is why the *maria* are round. The bright-white *terrae* are the ancient remnants of the Lunar crust that formed billions of years before the *maria*, and they have been subject to countless impacts by rocks large and small hitting the Lunar surface at hypersonic speeds. The *terrae* are bright white because they are composed chiefly of the white mineral anorthite.

Allan Hills 81005 houses fragments of white *terrae* and dark-black *mare*, rocks that formed in different Lunar provinces, billions of years apart. But how was it that the basalt and anorthite fragments came to be part of a single rock? Part of the answer lies in the third component of this marvellous meteorite: glass.

As red-hot liquid rock cools, minerals form, and the speed at which it cools is recorded by the size of the newly grown crystals. Slow-cooling liquid forms large crystals; fast-cooling liquid forms small crystals. If a molten rock is cooled rapidly enough it quenches – almost instantaneously transitioning from a liquid state into a solid – and the atoms do not have time enough to get themselves in order and slot neatly into an organised mineral structure. They remain a chaotic jumble of disorder, a frozen snapshot of their molten state. This highly disordered array of atoms is what geologists call a glass. Rapid cooling in the natural world only happens in abnormal circumstances, and so glass in the natural world is something of a rarity. It turns out, however, that glass is commonplace on the Moon.

The glass in Allan Hills 81005 formed when a celestial stone struck the Lunar surface at speeds many times that of a bullet leaving a gun. Hypersonic impacts impart enormous amounts of energy into the surfaces they strike, and, providing they are big enough, have the power to shock-heat solid rock into liquid in a fraction of a second. The Moon, lacking a protective atmosphere to filter out and slow down the occasional stray stone that crosses its path, bears the brunt of these impacts. When Allan Hills 81005 was assembled, veins of liquid rock at over 1,000 °C were injected around shattered fragments of *terrae* and *mare* like red-hot fingers, and upon instantly cooling they quenched to form the glass we see today. Written within this single, walnut-sized meteorite are billions of years of chaos and devastation played out on the Lunar surface.

Fragments of *terrae* and *mare* within Allan Hills 81005 formed in a similar way to the glass. Upon impact, much of the rock is not heated enough to melt but instead is shattered into fragmented shards and shrapnel ranging in size from burly boulders to pulverised powder. Vast curtains of rock are blasted into the Lunar sky and fall back to the surface as a chaotic blanket, sometimes thousands of miles away from the crater whence they came. This mixes different parts of the Moon together. The anorthite fragments from the *terrae* are mixed with the basalt of the *mare*, and the basaltic fragments from the *mare* are mixed with the *terrae*, and so on. When rocks are hit that hard they undergo 'shock metamorphism', where they are subject to pressures many hundreds of thousands of times the atmospheric pressure here on the surface of the Earth. In tiny fractions of a second, the mineral structures within the rocks are changed and the immense pressures are recorded within them. The entire surface of the Moon is pockmarked with impact craters, from the gigantic to the microscopic, and from time to time these impacts eject a piece of rock off the Moon's surface entirely. One such rock, by chance, would fall to the Earth and land on the East Antarctic Ice Sheet, and humans would name it Allan Hills 81005.

Over 350 kilograms of rock were picked up on the Moon and delivered to Earth by the *Apollo* astronauts between 1969 and 1972, and a few hundred grams were scooped up by the Soviet Union's *Luna* robotic landers between 1970 and 1976. Rocks brought directly back from the Moon by humans and robots are scientifically valuable for two main reasons: they have not been subject to degradation brought about by weathering, and we know exactly where on the Lunar surface they originated.

Pieces of the Moon delivered to Earth as meteorites, however, have lost all spatial context. To escape the Moon's gravitational field they must have been ejected from a crater at least a few kilometres across, of which there are hundreds of thousands sporadically pitting the Lunar surface. A particular Lunar meteorite could have come from any one of them.

There are (at the time of writing) just over 400 known Lunar meteorites. One, named Yamato 791197, was actually discovered in Antarctica three years prior to the discovery of Allan Hills 81005, but was not recognised as a piece of the Moon until after Allan Hills 81005 was identified.

As I write this, every single known meteorite from the Moon is a find. Not one has ever been witnessed falling from the sky.

Cosmochemists are eagerly awaiting the first Lunar fall, so be vigilant the next time you are looking up at our closest celestial neighbour on a dark night; you may witness a piece of it falling to Earth as a meteorite.

Rocky debris is ejected from the surface of an asteroid in the same way it is ejected from the surface of the Moon: impacts. Being tiny compared to planets and Moons, asteroids have weak gravitational fields, and so it does not take a collision of epic proportions to knock material off their surfaces. If fragments find themselves in a Kirkwood gap or happen to be ejected into space into just the right orbit, their traverse through interplanetary space may eventually bring them to Earth.

No human has ever visited an asteroid, but we have sent spacecraft to explore them on our behalf. So far we have encountered seventeen different asteroids up close, from the minute Itokawa (~ 500 metres across, about the same size as twenty-five double-decker buses parked end to end) to the largest one, Ceres (considered by some to be a 'dwarf-planet'), and returned detailed photographs of their surfaces.

As I write this book, the Japan Aerospace Exploration Agency (JAXA) *Hayabusa2* and the NASA *OSIRIS-REx* missions are both exploring two different asteroids: *Hayabusa2* has already scooped up pieces of asteroid Ryugu and will return them to Earth in late 2020; *OSIRIS-REx* will return pristine pieces of asteroid Bennu to Earth in late 2023. Both missions have acquired up-close images of their asteroids' surfaces. The astronomers of the nineteenth century only ever knew asteroids as faint 'star-like' points of light that moved across the sky, and had no way of knowing what they look like up close. They never knew how beautiful they are. We, who are lucky to be at the point in human history where the exploration of our Solar System has commenced, have seen a few of them in all their glory.

Asteroids come in a host of shapes, from the almost round, to the peanut shaped, to the potato-like, to the oddly square. They come in different colours, too: some are grey and are as reflective as bright dry sand on a beach, while others are dark and reflect about as much sunlight as a lump of coal. Some have uniform surfaces of constant brightness, while others have terrain that ranges from pitch black to as bright as freshly fallen snow. They differ enormously and come in many types, reflecting a rich tapestry of geologically diverse worlds. But they all have one thing in common: they are absolutely covered with impact craters.

Like the craters on the Moon, the size of craters on asteroids span from the gargantuan to the microscopic. In 2012, NASA's *Dawn* spacecraft imaged the giant Rheasilvia impact crater on the surface of the second largest asteroid, Vesta. At over 500 kilometres across (about half the length of Great Britain) and around twenty kilometres deep, Rheasilvia almost completely obliterated Vesta when it formed, ejecting about six million million million kilograms of rock into interplanetary space. In the centre of Rheasilvia and towering twenty-five kilometres above the crater floor is the tallest known mountain in the Solar System, which formed as the ground rebounded like a raindrop splashing into a puddle.

The countless number of impact craters peppering the surfaces of the asteroids provide ample opportunity for material to be blasted away from their surfaces and into space. Impacts leave their mark written inside the meteorites themselves too and, as with Allan Hills 81005, traces of catastrophe are laid bare by the glass. Havoc is wreaked on the minerals within: once beautifully shaped crystals are riddled with fractures; their crystal structures are warped on the atomic scale; and they are marred by partial melting. In some cases, when the atoms are rearranged by the immense pressure of an impact, minerals transform into new and exotic versions of their former selves. Diamonds, jarred into existence faster than the flash of a camera, are amongst them, and decorate some of the most severely impacted meteorites.

Cosmic stopwatches

Once a meteorite has been ejected from its parent body, whether it be an asteroid or the Moon, its journey through interplanetary space to the Earth can be timed using natural isotopic stopwatches. There is a constant flow of high energy atomic particles streaming through our Solar System called 'cosmic rays'. Sent on a journey through interstellar space by the explosive deaths of distant stars, these particles travel at speeds approaching that of light. If one happens to hit a solid object – an asteroid, for example – the cosmic ray will meet its end: after travelling hundreds of light years through the galaxy, it can be stopped dead in its tracks by a centimetre or two of rock. The rock does not escape unscathed, however.

Most cosmic rays are made of single protons, and even though they are tiny and have hardly any mass, their extraordinary speed means they can pack a punch. Like atomic sledgehammers, they can spark nuclear reactions on collision with the atoms making up solid rock to produce characteristic isotopes of certain elements called 'cosmogenic nuclides'. The longer a rock is exposed to the steady stream of cosmic rays passing through our Solar System, the more cosmogenic nuclides are produced within it. When exposed to space, their numbers increase with the passing millennia.

Since most of the rock making up an asteroid is hidden below the surface it is shielded from the pitter-patter of cosmic rays, and never has the chance to accumulate cosmogenic nuclides. But once a piece of rock is liberated from the depths of its parent asteroid by an impact event and ejected into interplanetary space, it loses its rocky shield and is exposed to the full vigour of the cosmic rays. Cosmogenic nuclides begin to accumulate within the rock like the sand that accumulates in the bottom of an hourglass; the stopwatch begins ticking. All the while the stray rock is travelling through space, it lacks protection from cosmic rays, and so the cosmogenic nuclides steadily pile up. The longer it is in space, the more they accumulate.

Upon falling to the Earth's surface, the meteorite gains protection once more from the battering of cosmic rays. The Earth's thick nitrogen atmosphere and the giant magnetic field emanating from the planet's centre filter out most cosmic rays, preventing them from reaching the surface; the production of cosmogenic nuclides within the meteorite ceases. The clock stops ticking.

We can calculate the rate at which cosmogenic nuclides are produced in celestial rocks during their transit time to Earth by measuring the intensity of the stream of cosmic rays passing through our Solar System; that is, we can calculate the rate at which the isotopic stopwatch ticks. By measuring the extent to which cosmogenic nuclides have accumulated in a meteorite in a laboratory we can calculate how many times the stopwatch has ticked; this tells us how long the meteorite took to reach the Earth from its parent asteroid.

Given that meteorites do not travel in straight lines – they go into orbits around the Sun, which happen to cross the Earth's orbit – it takes a long time for them to reach us. Whilst some stony meteorites arrive on Earth in little more than 100,000 years, most have a transit time of between ten and thirty million years. Many iron meteorites took their time, however, and were orbiting the Sun for twenty times longer than most stones: some spent up to 500 million years or more in interplanetary space before reaching the Earth, allowing time for vast quantities of cosmogenic nuclides to accumulate within them.

Exactly what causes the huge difference in transit time between the stones and the irons is one of many mysteries held by meteorites, but one possibility is that iron meteorites simply survive longer in interplanetary space. Being made of stern stuff, lumps of metal can survive long term in space without being chafed away and turned into dust, whereas the fragile stony meteorites are easily eroded by small collisions on their way to Earth. They are more easily timeworn. There is probably a limit to how long a stony meteorite can survive in outer space once it has been ejected from its parent asteroid, which is why none of them record a vast transit time on their cosmic ray stopwatches.

Like the Earth, the Moon, and the other planets, the asteroids were hot when they formed. But, having never coalesced to form a planet-sized body, they did not incorporate large quantities of radioactive isotopes – the fuel that powers planetary-scale heating – into their rocky bulks. Being minute in comparison to planets, the little radioactive fuel that asteroids did assimilate when they formed decayed away into insignificance in no time (geologically speaking), leaving them to lose what little internal heat they did have to the briskness of space, and cool off.

The laws of nature are the same everywhere, and what is true for a cup of tea here on Earth is true for an asteroid orbiting the Sun between Mars and Jupiter: small things cool down faster than big things. The reason for this is elegantly simple: it is due to the surface area to volume ratio. Being (roughly) spherical, the volume of a planet (or an asteroid) increases with the cube of its radius, but its surface area only increases with its square.[5] Thus, the larger a celestial body gets, the more heat it contains in its bulk and the less efficiently it loses that heat to space through its comparatively smaller surface area, and so it stays warmer for longer.

Being minute compared to planets, the asteroids lost their heat incredibly quickly, and even the largest were stone-cold after a few tens of millions of years. To our human minds this seems like a long time, but when it comes to meteorites we must onsider them against the backdrop of geological time. In our timeline where we squeezed the entirety of Earth's history down into a twenty-four-hour day, the asteroids were cold in little over half an hour, and they have remained frigid ever since. The Earth, by comparison, is something like 4.6 billion years old but is still producing enough heat to send rivers of red-hot liquid rock to the surface on a daily basis.

Having chilled so soon after they formed, asteroids have long been geologically dead worlds, and so the rocks from which they are made have remained entirely unchanged – aside from the shock of an occasional impact – since their formation. This allows these rocks – which fall to the Earth as meteorites – to take us back in time further than any rock on the Earth. They can tell us things no Earth rock ever could.

Of all the tales told by the stones that fall from the sky, perhaps the grandest and most astonishing is the assembly and coming together of our cosmic home: the Solar System.

3

FROM GAS TO DUST; FROM DUST TO WORLDS

As the place where each of us lives out our lives, and as the home planet of the human species, we place a great importance on the Earth. It's history is intimately entwined with our own.

A deep longing stirs inside us when we see photographs of our blue ocean world taken from outer space. It is our home. Our ancestors felt the same deep connection with the Earth. Questions about how our home planet came to be and how we came to walk upon it were first asked in antiquity, and every culture and religion across the planet has come up with their own Creation myth in an attempt to answer these questions. So far as we know, the Earth is the only place in the Universe where molecules evolved into minds and can question their own origins; and we, humans, so far as we know, are the only pieces of the Universe able to find anything resembling an answer. To wonder where we came from is an intrinsic part of what it is to be human. The question is inside every one of us.

Each of our bodies is made from some ten million million million billion atoms that come from the air we breathe, the water we drink, and the food we eat. These in turn come from the Earth's oxygen-rich atmosphere, the water that flows across the surface and rains from the sky, and, in the case of our food, from plants that grow in the ground from the gases in the air. Every time you eat, drink, or breathe, some of the atoms that enter your body from the 'outside' stay and, temporarily at least, go on to make new cells and become part of 'you' on the inside. We are in a real sense made of Earth. The questions of how the Earth came to be and how we came to be here are, if you go back in time far enough, one and the same.

We are living in an era of human history where the true extent of our planetary and celestial origins are at long last coming to light after 200 millennia of wondering. Meteorites played a pivotal role in this great awakening.

Nebulosity

It did not escape the notice of the early astronomers that everything in the Solar System, aside from the occasional oddity, appears to spin the same way. All of the planets, comets, and asteroids orbit the Sun in the same clock-like direction; the vast majority of the moons in our Solar System orbit their host planets in this same direction; six of the eight planets spin like tops in this same direction, too, as do almost all of their moons.* Even the Sun, the beating stellar heart of our Solar System, rotates once every twenty-five days in the same direction as the planets orbit and spin. It is as though the Solar System is caught up in a whirling and everlasting current.

The Solar System is incredibly flat, too, and when viewed edge on, the planets and major moons orbit the Sun within an

---

* Venus, the second planet from the Sun, is a noteworthy oddball: it rotates the 'wrong' way. Exactly why it spins in the opposite direction to most planetary bodies in the Solar System is a mystery. Uranus, the ice giant, which is the seventh planet from the Sun, orbits almost perfectly on its side, probably due to a massive collision knocking it sideways early on in the Solar System's history.

exceedingly tight plane, as if racing around concentric grooves on an invisible table top; if one shrank the Solar System down such that the orbital path traced by Neptune – the planet furthest from the Sun – was twenty centimetres across, the Solar System would be about as flat as a seven-inch vinyl record. Only the asteroids and comets, which represent little more than orbiting debris when compared to planets, deviate significantly from this narrow orbital plane; asteroids orbit the Sun on inclined planes tilting as much as forty-five degrees, and some cometary orbits are inclined at full right angles.

The first detailed attempt at explaining the vortex-like nature of the Solar System was undertaken by German philosopher Immanuel Kant in his 1755 book *Allgemeine Naturgeschichte* (Universal Natural History). It was a marvellous, and accurate, stroke of speculation. Kant proposed that immediately after Creation, the Solar System hung in space as a vast cloud that was dispersed as a formless chaos.

Kant hypothesised that the nebulous cloud was slowly turning in space, and with a nod towards the Newtonian view of gravity, he suggested that it began collecting together into a dense thicket, spurred onwards by its own gravitational pull. As it collapsed further, its gravitational pull increased, and so it collapsed further still. Propelled by its initial turning motion, the pivoting continued as the thicket grew tighter and denser, eventually somehow transforming into the rotating Sun. He went on to describe how some of the material in the cloud was deflected sideways and away from its free fall into the Sun, falling into circular orbits around it along a common flattened plane; this residual material, he said, formed numerous smaller clumps which themselves collected together to form the planets.

In 1796, celebrated French mathematician Pierre-Simon Laplace presented his own ideas about how the system of planets – including the Earth – and comets orbiting our Sun formed from an initially formless shroud of gas that was slowly rotating in space. Laplace described how the rotating cloud collapsed under the crushing force of its own gravitational field, spinning ever faster as it did so, to form a giant sphere of dense gas that ignited to become the Sun. He envisaged great swathes of material tearing away from the rapidly pirouetting Sun and being thrown off into space in the form of concentric rings, each in orbit around the Sun. The new planets, he said, formed from these swirling rings around the Sun and continue swirling onwards in the same orbits to this day.

Our current view of the Solar System's formation is, by my estimation, more wonderful and more profound than any myth of old. As we will discover, it involves the explosive deaths of giant stars, stellar winds tearing through interstellar space, and the undoing of worlds and the forging of others. Kant and Laplace's ideas, while lacking in detail, turned out to be right in essence.

At this point it is worth taking a moment to orient ourselves in the landscape of deep time. The Universe – that is, everything in existence – originated in the Big Bang just shy of fourteen billion years ago. Our Solar System came into existence just over four and a half billion years ago, which is almost exactly one-third the age of the Universe. It is a story with a grand opening.

Flickering to light

In the beginning, our Solar System was part of a bitterly cold cloud of formless gas, sparsely sprinkled with minute grains of rocky dust. At first incredibly thin, the interstellar cloud would have been comprised of mostly gaseous hydrogen and helium; an Earth-sized sphere of such haze would only contain a few kilograms of matter, which is about as much mass as a newborn baby. It was so thin it was almost not there at all. Such clouds we see draped across the night sky today, and we appropriately name them after the Latin word for mist: 'nebula'. Despite being little more than slight wisps of gas with smatterings of dust, nebulae span immense distances across interstellar space, stretching onwards for tens to hundreds of light years.*

Nebulae are visible amongst the sea of stars that embellish our night skies, and with ill-defined shapes and edges that dissolve away into nothingness, they appear as faint smudges against the black backdrop of space. A few nebulae are visible to the unaided eye even in places where the sky is bleached orange by stray street-lamp light: the skies of the winter months in the Northern Hemisphere are graced by the nebulae in the constellation of Orion the Hunter (the nebulae are in his sword) and the faint smudge that seems to surround the Pleiades star cluster. These two nebulae are particularly easy to spot, and are an astronomical wonder to behold for even some light-polluted city-dwellers.

When we look at them through powerful telescopes, nebulae are transformed. Intertwining filaments of gas weave around one another; columns of dense clouds tower like chimneys; ethereal wisps are drawn outwards into the cosmos as long fila-ments. Some nebulae are pitch black and appear as inky blotches hanging in the sky through which no stars are visible, but most glow with a dim red hue. Far too cold to shine with their own light, they are irradiated by the light from nearby stars, causing their gas to glow incandescent. Minute transitions of energy on

---

* One light year is the distance travelled by light in one year and is equal to nine and a half trillion kilometres.

the subatomic scale within the gas add up across a cloud that spans galactic distances, and send nebulashine pouring outwards into the cosmos.

Winds from particularly hot and energetic stars drive vast swathes of nebulae backwards and carve out large regions of empty space, sculpting tendrils of dense gas at the peripheries. The explosive deaths of stars many times the mass of our own Sun send forth powerful shockwaves that traverse interstellar space and ripple outwards through the nebulae, bunching together gas into threads of high-density ripples as they go. Massive stars draw immense strings of nebulae around themselves with their strong gravitational pull, their tides drawing together thinly dispersed gas into patches of thick fog. Interstellar clouds, whilst static on a human timescale, are dynamic places that heave and flow as the aeons ebb by.

Nebulae are kept aloft by turbulent currents and strong magnetic fields within the cloud, causing them to slowly billow and whirl. In regions where nebulae are brought together in pockets of high density by stellar influences, gravity takes hold and nebula collapse ensues. Gas and dust are drawn inwards under the influence of gravitational attraction to produce nodules of high-density nebula that trigger a runaway process of subsidence. The denser the core of a nodule becomes, the stronger its gravitational pull on the surrounding nebula, therefore enabling it to pull in more of the surrounding gas and dust, and increase in density. Once a nebula has reached this point there is no turning back. Many hundreds of these dense cores may form from a single nebula, each drawing in gas and dust. We see the process of nebula collapse taking place in real time today using powerful telescopes, and in the Orion Nebula alone over 200 pockets of collapsing nebula have been catalogued, each one representing a nodule of nebula giving way to the force of gravity.

Many thousands have been catalogued elsewhere across the sky.

Our own Solar System was once like this too. A small part of the bitingly cold cloud from which it formed began collapsing inwards somewhere just over four and a half billion years ago, creating a nodule of high-density gas and dust billions of kilometres across. At first it was pitch black inside the nodule, because starlight could not penetrate the opaque gas and dust. As the core of the nodule drew in more surrounding gas, it slowly grew denser and more massive, and the temperature began to slowly rise. As gravity took a tighter grip on the growing core, the temperature rose past the thousands of degrees upwards into the millions.

Then it flickered to life.

Nuclear reactions – the fusion of hydrogen into helium – initiated by searing temperatures and crushing pressures ensued in the core of the crumpling nebula, flooding our nascent Solar System with starlight for the first time. This moment marked the birth of the Sun: the age of light in our Solar System's long history had begun.

With fresh starlight shining outwards from the centre of the collapsing nebula, what remained of the gas and dust was illuminated for the first time. A flat disc of gas, tens of billions of kilometres across – called a 'protoplanetary disc' – had by now formed around the newborn Sun and was orbiting it like a merry-go-round.

Before the onset of collapse, our natal nebula was slowly rotating, and this initial rotation was maintained and amplified throughout the gaseous and dusty infall by a fundamental law of physics: the conservation of angular momentum, which, in its simplest form, says that spinning objects will continue to spin unless acted upon by an external influence. With nothing to halt the spinning motion of the nebula during its collapse, it continued to pivot around its centre, and as it contracted, the spinning grew faster. Just as a ballerina pirouetting *en pointe* spins faster as she tucks her leg closer to her body.*

In Nature, spinning things tend to want to flatten out. Before the onset of collapse, the gas particles and the minute motes of dust were moving around in turbulence, and while overall the nebula was rotating, the motion of individual particles was essentially random. As the gas and dust were drawn together by the influence of the gravity from the core, their up-and-down and their side-to-side motions cancelled out, causing them to form a flat disc shape. The disc, in orbit around the massive star at its centre, kept its rotational motion thanks to the conservation of angular momentum. The celestial pirouette continues to this day.

Then came the rocks

From an initially shapeless interstellar cloud, Nature brought about order. It took little more than a few million years for the Sun and the protoplanetary disc to form from the collapsing nebula, which on a geological timescale is practically instantaneous. In our twenty-four-hour geological day, it look little over ninety seconds, and so after the first minute and a half, the Solar System had practically finished forming.

While the Solar System was being drawn inwards by gravity and flattened by the conservation of angular momentum, other stars were forming in close proximity to our own. Some of them were stellar giants who had astonishingly high masses – some

---

* Another demonstration of the conservation of angular momentum can be achieved using a simple office chair. If you spin around on an office chair and stick your arms and legs out you will spin slowly; tuck your arms and legs in and you will spin faster.

stars reach ten or fifty or even 100 times the mass of our own Sun – and as a consequence, burned through their fuel incredibly quickly: in the time it took our Sun and protoplanetary disc to come together, they were spent. Ferocious stellar winds emanated from their surfaces as they consumed the last dregs, and streams of heavy elements, forged deep inside them by nuclear reactions, were blown outwards into the surrounding nebula. Then the stars switched off. No longer able to hold themselves up against their own enormous gravitational fields, each catastrophically collapsed before immediately exploding as a supernova.

The chemical cocktail of heavy elements cooked up in their astonishingly hot and dense cores seeded the surrounding nebulae with newly forged elements. Some of these freshly synthesised elements were sprinkled onto our own protoplanetary disc as it was flattening. As nearby supernovae popped like cosmic firecrackers, their shockwaves tore through our newly formed Solar System.

Vast amounts of energy were released by the collapsing nebula, and so the once-frigid cloud became blisteringly hot. Enough energy was released not only to melt, but to vaporise most rocks and, as a result, most of the rocky dust caught up in the collapsing nebula was turned into gas. The temperature of our Solar System at this point in its history was beyond ordinary comprehension, and from the outside would have resembled an angry-looking disc of glowing gas circling our young star.

As the disc orbited the infant Sun and grew denser, the gas began to change. The atoms began to communicate chemically as they were packed into a progressively smaller and smaller volume. The disc was cooling, too, as it radiated energy away into interstellar space. Within 100,000 years or so, minerals – solid pieces of matter – began to form. Geology, at last, was born from the wisps of nebulosity circling the young star.

Here on the Earth we are used to minerals forming from liquids, usually as red-hot lava cooling to form igneous rocks or as salts precipitating out of mineral-laden fluids, but in the protoplanetary disc, minerals formed directly from the cooling gas in a process called 'condensation'. Near the Sun where temperatures were scorching – well in excess of 1,500 °C – the only minerals capable of condensing were exotic oxides of aluminium and calcium (minerals containing different proportions of oxygen, aluminium, and calcium). Nothing else is chemically stable at such searing temperatures. The first mineral to form from the glowing hot gas as it slowly cooled was corundum, known chemically as Al_2O_3 (aluminium oxide), and colloquially as ruby or sapphire.[*] From gas to dust.

Sparkling as it orbited the young Sun in the disc, the condensation of corundum marked the first time a solid piece of our Solar System had formed, but as the gas cooled further, it reacted with the corundum and consumed it. But the destruction of one mineral leads to the formation of another, and from the corundum came a new mineral called hibonite ($CaAl_{12}O_{19}$). Like corundum, pure hibonite is colourless, but occasional impurities will contaminate the newly condensed crystals and render them vivid blues, deep greens, or ripe oranges. Further cooling allowed the gas to condense a sequence of new exotic minerals rich in calcium and aluminium: perovskite ($CaTiO_3$), followed by melilite ($Ca_2Al_2SiO_7$), followed by spinel ($MgAl_2O_4$). The unusual environment produced unusual geology.

---

[*] The pure form of corundum (Al_2O_3) is colourless. Ruby and sapphire obtain their colours from chemical impurities: ruby is aluminium oxide containing tiny amounts of chromium, and sapphire is aluminium oxide containing tiny amounts of other metals such as titanium or iron.

Onwards the gas in the great disc cooled as it orbited the Sun, slowly condensing and crystallising a sequence of minerals one after the other. Most were no larger than a mote of dust.

Further away from the broiling solar surface and further outwards into the protoplanetary disc, the temperatures fell low enough to allow condensation of olivine dust (peridot) and blobs of metallic iron. Still further away from the Sun, dusty flecks of a mineral named feldspar condensed, joining the host of other minerals that had formed from the gas already. Thus, the first rocks in the Solar System were forged and the rock record began.

Half a billion kilometres away from the Sun in the depths of the slowly cooling disc (around twice the distance from the Sun to Mars), the temperature plummeted low enough for flecks of water-ice to condense. The imaginary line encircling the Sun outside which the temperature is frosty enough for water-ice to form is affectionately known to cosmochemists as the 'snow line'. If an icy flake drifted Sunwards over this line, it evaporated and became gas once more. Far beyond the snow line where the temperatures plunged even lower, more flakes of ice condensed, but not of a watery nature – shards of ammonia-ice and methane-ice formed from the wintery gas. Out in the bitterly cold reaches of our young Solar System, the newly formed ices were lit by a distant, much fainter and colder, Sun.

Amongst the variety of ices forming in the disc, a complex array of organic molecules were spontaneously synthesised from a series of intricate chemical reactions. Organic molecules are a class of chemical compounds that contain carbon in their molecular structures. They are chemically elegant, often with many molecular arms capable of grabbing and reacting with surrounding atoms. They are also the molecules from which life originated. The rock record is not the only story to find its beginnings in the protoplanetary disc – the story of life may have too. But more on that in a later chapter.

World building

Powerful stellar winds flowing forth from the Sun blew the dust backwards into the colder reaches of the disc along soaring arcs. Mixtures of high-temperature condensates and low-temperature condensates came together in dust-rich regions of the disc, piling up to form concentric lanes orbiting the Sun. Turbulent winds and eddies in the thin wisps of gas further concentrated dust into whirling clouds like cosmic tumbleweeds.

As the condensed minerals orbited the Sun, the largest no bigger than the nail on your little finger, the motes of dust began to interact. Static electricity drew the specks together into tiny clumps, like the dust bunnies you might find underneath a neglected piece of furniture. If two grains collided too quickly they would ricochet off one another or splinter into smaller fragments. High-speed collisions between particles produced shards of crystalline shrapnel, but many collisions were gentle enough to culminate in sticking. Large grains stuck to minute flakes to produce a chaotic mixture with a mishmash of grain sizes. In the colder reaches of the disc, sticking was enhanced by the presence of ices, whose pliancy allowed them to act like cohesive glues.

With each orbit around the Sun, the horde of clusters grew and grew, becoming more compacted upon each collision. Wafted into dense tussocks by turbulent nebula winds, large clouds of these clusters accompanied by innumerable motes of finer dust accumulated to the point where a familiar force took hold: gravity. Giving way to their mutual gravitational attraction, the massed clusters coalesced to form compact aggregates of dust kilometres across. Planetesimals – the first rocky bodies and the progenitors of the planets – were born. From dust to worlds.

A proliferation of rocky planetesimals emerged, each anywhere between one kilometre and 100 kilometres across.

Firmed up by an untold number of minuscule collisions with dust-sized grains, and compacted by their own gravitational fields, the planetesimals had the first solid surfaces in the Solar System; despite their weak gravitational fields you could in principle stand on one. Those that formed in the scorching inner regions of the Solar System were bone dry, assembled exclusively from rocky material. Beyond the snow line, planetesimals assimilated ices and organics, turning them into freezing cold worlds of sludge and hydrous minerals. As the planetesimal swarmed around the Sun, each interacted with the other through mutual gravitational attraction and their orbits began to evolve.

Gravity shaped the physique of these newly formed worlds, too. Those that collapsed from larger, denser clouds of dust assimilated more material into their rocky bulks, giving them a strong gravitational pull. If a planetesimal grew to more than 250 kilometres – the 'potato radius' – it transitioned from a lumpy pile of rubble into a spherical rocky ball, resembling a miniature planet.

For some planetesimals, their story ended before they really began. Near misses as they careened past one another sent their trajectories haywire, and many met their end by plunging into the Sun. Some met a far colder fate and were ejected out of the Solar System entirely, slingshotted into interstellar space.

Near misses nudged the trajectories of many planetesimals towards each other in a process called 'gravitational focusing', increasing the chance of a future collision, and, with so many solitary worldlets circling the Sun, clashes were inevitable. Collisions often spelt a catastrophic end to a planetesimal's brief tenure in the Solar System, but some fragments went on to be swept up by other planetesimals and became part of a brand-new infant world. Creation; destruction; renewal.

Gentle mergers

Despite the astonishing speeds at which the planetesimals hurtled around the Sun – typically upwards of tens of kilometres each second* – not all collisions ended in catastrophe. If two planetesimals followed similar orbital paths it was almost inevitable that at some point they would find themselves in the same place in space at the same time, and collide. However, their low speed relative to one another meant that gentle merging could ensue, and the planetesimals stuck together to form a single larger body. Larger planetesimals, with higher masses, efficiently directed smaller bodies into their orbital path through gravitational focusing; and they quickly snowballed in size. With each orbit around the Sun, the lucky few gobbled up smaller planetesimals, growing into 'planetary embryos'.

Not all of them were destined to develop into fully fledged planets, however. Many planetary embryos were destroyed in violent collisions, and some of them spiralled into the Sun. The fledgling planets that did survive managed to avoid such fates, and they lasted long enough to establish themselves in their own stable orbital lane. Orbiting in widely separated concentric paths, the fortunate few planetary embryos swept up the remaining planetesimals and dust. Carving out lanes of empty space within the disc as they orbited, they slowly cleared away the rocky debris from the Solar System.

Their final stages of growth were marked by impacts and mergers on an apocalyptic scale, but by this point the planetesimals were too big to be fully disrupted. Surviving annihilation and maturing into wandering giants many thousands of kilometres across, they became fully fledged planets.

---

* At this sort of speed, it would take little more than one and a half minutes to travel in a straight line from John o' Groats to Land's End.

All that is now Earth was once sky. The solid ground beneath our feet, an entire planet's worth of rock, was assembled from the coming together of innumerable motes of microscopic nebula dust.

Each planet, forming at different distances from the Sun, grew from a unique blend of planetesimals and dust, thus attaining its own unique chemical and isotopic composition. The first four planets from the Sun grew entirely from rocky material (some with smatterings of ices) to form the inner Solar System: Mercury the scorched sphere, Venus the morning star, Earth the blue marble, and Mars the red planet. Further away from the Sun, the planets grew from a combination of rocky material, ices, and gas: Jupiter the giant, Saturn the graceful, Uranus the ice giant, and Neptune the bitterly cold. Beyond the orbit of Neptune, countless icy bodies (including Pluto and Charon) orbit the Sun. The lopsided rotations of Uranus may have been caused by the cataclysmic impacts of planetesimals on the newly formed planet; such impacts may also be the reason Venus spins in the wrong direction. The unique composition and character of each planet – both chemical and isotopic – is a relic from their unique blend of dusty building blocks.

The myriad of moons in our Solar System – over 150 currently known – formed in a multitude of different ways, causing them to be as geologically varied as the planets. In the same way the residual dust and gas that survived falling into the Sun went on to form the planets, the dust that survived falling into the planets went on to form moons.

At least one moon had a far more violent origin: ours. *The Moon* formed when a planetary embryo cataclysmically smashed into the Earth shortly after it formed. The impact was glancing rather than head on, but even so, a great curtain of vaporised and liquidised rock was blasted from the surface of the young planet into space. Much of this material immediately rained back down onto the molten surface, but some stayed in space and coalesced to form the Moon.

Within around fifty million years after the formation of the first condensates, the Solar System had run out of planetary building blocks. The era of planet formation was over. From a common origin, the stories of the eight planets – four rocky worlds and four gaseous worlds – and the myriad of moons and smaller worlds diverged, and each followed their own unique path into the deep future.

In time these worlds would evolve and change almost beyond recognition as deep time took its toll: active volcanoes would coat the surfaces of at least four planets; liquid water would flow across the rocky surface of at least two; many planets and moons would acquire atmospheres from the gas they coalesced; the gas planets would each develop their own exquisite array of concentric rings. At least one planet, of course, would go on to harbour life.

Asteroids

By a quirk of gravitational whimsy and celestial arrangement, the planetesimals and planetary embryos in some parts of the disc were prevented from clumping together by the massive gravitational pull of Jupiter (and, to a lesser extent, Saturn). The particular arrangement of the gas giants, which formed not long after the Sun, saw that the planetesimals caught in these regions forever remained a band of isolated wanderers. Planets did not form in these parts of the Solar System, and the planetary building blocks remained a disassembled mass orbiting the Sun.

Regions of disassembly existed both inside and outside the orbit of Jupiter, which lies around five times further away from the Sun than does the Earth. The disassembled planetesimals, having formed from different blends of condensed dust, formed two distinct populations: one, inside Jupiter's orbit, was largely rocky; whereas the other, forming outside Jupiter's orbit, where it was colder, was a combination of rock and icy minerals. The gas giant's orbit wobbled back and forth numerous times as the Solar System was stabilising from its nebulous inception and, as it migrated, the two populations of planetesimals were scattered far and wide, destroying many of the nomadic worldlets in the process. But some survived. There was an exodus of ice-rich planetesimals from the cold outer Solar System, and, upon being scattered into the tepid inner Solar System, they slotted in and amongst the population of rocky worldlets that dwelled there already. The inhabitants of this part of the Solar System, prevented from clumping together by the gravitational pull of Jupiter, survived for over four and a half billion years. These surviving planetesimals are the asteroids in the Asteroid Belt.

The Asteroid Belt is populated by millions of asteroids, and they formed in many different parts of the Solar System. Today, there are 'only' 3,000 million billion tonnes of rock left in the Asteroid Belt, which, at 0.05 per cent of the total mass of planet Earth, would not be enough to make a planet even if they did coalesce. (Mercury, the smallest rocky planet, contains around 100 times more rock than the entire Asteroid Belt put together.)

Many of the ice-rich planetesimals from the outer Solar System were scattered into highly elliptical (elongated) orbits around the Sun. Today we know them as the comets, and as they careen past the Sun, great streams of ices and organic molecules evaporate from their surfaces and billow into space. Comets, which are made largely of ices, represent one extreme end of planetesimal character, and asteroids, assembled largely from rocky materials, represent the other. In reality, most planetesimals lie somewhere in the middle: even the most ice-rich comets contain pieces of rocky dust, and even the rockiest of asteroids show faint traces of the tell-tale signs of ancient water.

Time was as important as location in affecting the character of an asteroid. As the first motes of dust were condensing from the cooling protoplanetary disc, the emerging Solar System was awash with rapidly decaying radioactive isotopes (also known as 'radioisotopes'). Blown into the collapsing nebulae from the atmospheres of nearby giant stars, the short-lived radioisotopes rapidly decayed and liberated vast amounts of nuclear energy over a short period of time; they lived fast and died young.

One of the liveliest radioisotopes present in the early Solar System was aluminium-26 (^{26}Al), which decayed away into insignificance in little over three and a half million years, releasing huge quantities of nuclear energy in the process. Planetesimals that formed early, before the ^{26}Al had decayed away, incorporated vast amounts of this nuclear fuel into their rocky bulks and therefore melted. This destroyed the nebula dust from which they coalesced. A short geological history, featuring plenty of red-hot liquidised rock, played out on these worlds before they quickly lost their heat to space and froze forever more.

The final dregs of dust to coalesce did so after the infant Solar System had cooled off. Within a few million years after the formation of the protoplanetary disc, most of the short-lived radioactive isotope had decayed away into nothingness, starving later planetesimals of the nuclear fuel they needed to melt. Whilst many of the planetesimal latecomers did warm up slightly, they remained relatively cold. They preserved the motes of nebula dust from which they formed. The asteroids, and the meteorites they send forth, are the custodians of nebula dust – pieces of our primordial Solar System.

When we gaze skywards through telescopes into the incandescent gas and newly forming planetary systems at the heart of the Orion Nebula, we are looking into our own deep past. Discs forming around infant stars; hot young stars carving out cathedral-like spaces from the thin wisps of nebulae; giant stars disseminating freshly synthesised elements into their surroundings. In other parts of the sky, astronomers have glimpsed concentric gaps in protoplanetary discs that were carved out as planetesimals sweep up the gas and dust. We are witnessing the formation of new planets; new places. We have learned much about how new systems of planets form and evolve using telescopes, but there remains a fundamental problem of distance; nebulae lie tens, hundreds, or thousands of light years away from the Earth. There is only so much we can learn by watching from afar.

Earth rocks can only take us so far back in time, but meteorites offer a way to get our hands on pieces of a newly forming Solar System. These pieces of asteroid shrapnel preserve the Solar System's earliest history, all the way back to the genesis of the rock record when the first dust condensed from the nebula and the subsequent forging of planetesimals.

The distinct geological character of meteorites reflects the two distinct types of asteroids: those that melted and those that did not. It is in this manner that we divide meteorites into two great families: those that originate from unmolten asteroids are named 'chondrites', and those that originate from molten asteroids are named 'achondrites'.

The achondrites record the short-lived geological evolution of the molten asteroids. With a nascent Solar System brimming with newly formed rocky worlds, many new stories were hastily written into these stones, and the tales they recount are just as strange and wonderful as the tales from the collapsing nebula that immediately preceded them.

4

SPHERES OF METAL AND MOLTEN STONE

Melting is a force of geological destruction. The rocks of the Earth are routinely subject to crushing tectonic pressures and unyielding weather, but nothing quite undoes them as effectively as heat. Upon transitioning from a solid state into that of a liquid, the atoms in a rock cease to be bound to one another by chemical forces, and fall to pieces on the atomic level. Almost the entire character of a rock is lost for ever during melting. Nature makes good use of molten rock, however, and, from it, crafts rocks anew.

Rocks from the Earth are not the only ones to have experienced the renewing effects of melting. Some of those that fall from the sky have, too. The radioactive isotope of aluminium – ^{26}Al – was a particularly potent heat source for planetesimals. As the radioisotopes rapidly decayed they swiftly released the atomic energy contained within their nuclei, melting many planetesimals in their entireties. The nebula dust from which these planetesimals coalesced was completely destroyed in the process, as whole worlds were transformed from agglomerations of dust into glowing orbs of liquid rock.

Most molten planetesimals, being small, were cool within a few million years or so. Their geological heat engines ran out of power quickly and they froze over. Even the largest – which held onto their internal heat for longer – were cold after something like 100 million years. As the liquid rock froze and crystallised, the molten planetesimals transformed into solid bodies once more, freezing within them stories of intense heat and complete chemical transformation. Achondrites, the meteorites that originate from these molten asteroids, are the oldest igneous rocks that we know of.

Small-scale changes – the transition of minute grains of dust into liquid rock – were just the beginning of the transformation. The entire internal structure of molten asteroids was completely upended in a process called 'differentiation', during which they transformed from dusty bodies that were more or less uniform throughout into bodies with two distinct geological layers: a metallic core, overlaid by a stony mantle with a thin outermost crust.

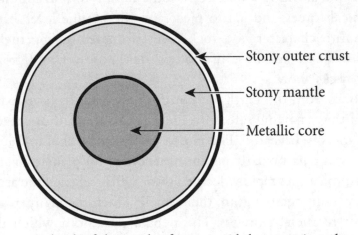

Stony outer crust

Stony mantle

Metallic core

A sketch of the inside of an asteroid that experienced extensive melting early on in its history. Although the Earth is many hundreds or thousands of times bigger than most asteroids, it would look similar if sliced in half.

An affinity for iron

One of the most abundant elements in the nebula dust was iron, and when the dust incorporated by planetesimals melted, it was liberated. Despite the gravitational field of the planetesimals being extraordinarily weak, it was enough to gently tug the unshackled iron to the centre of the body. The metallic iron sank slowly downwards through the searing magma due to its high density, and collected in vast amounts.

There is a suite of chemical elements that geologists call the 'siderophiles'. The word siderophile comes from the Ancient Greek σίδηρος (*sideros*) meaning 'iron', and φιλία (*philia*) meaning 'beloved': siderophile elements are 'iron loving'. They hold a high chemical affinity for iron. In geological systems – Earth bound and celestial alike – siderophile elements tend to follow iron as it moves from one mineral to another. Wherever iron goes, the siderophiles are hot on its heels. The fourteen siderophile elements include nickel, platinum, iridium, tungsten, and gold. When these elements were liberated from the nebula dust upon melting, they followed iron downwards into the centre of their parent planetesimal, pooling alongside it to form a vast blob of metallic magma, known as the core.

If one laid eyes on the molten metallic core of a recently differentiated planetesimal, it would be glowing a brilliant red hot with the radiance of sunlight. This core, however, would have been shielded from view by the overlying layer of less dense magma left behind by the sinking metal. Having been stripped of most of the iron and almost all of the siderophile elements, the chemical make-up of this outer layer changed to become vastly different from the sphere of metallic magma below.

An affinity for oxygen

The magma surrounding the metallic core at the centre of a molten asteroid was rich in elements that geologists call 'lithophile' elements, and these largely formed the layer directly above the core: the mantle. *Lith* comes from the Ancient Greek λίθος (*lithos*) meaning 'stone'. Lithophile elements are 'rock loving' and have a high chemical affinity for oxygen, eagerly combining to form an array of oxygen-rich minerals common in so many rocks. Prevented from sinking towards the centres of the molten planetesimals thanks to their low density, they remained in the upper layers, floating atop the molten metal in the core below. Lithophile elements are some of the most abundant in rocks here on the surface of the Earth and include the familiar silicon, aluminium, calcium, sodium, and magnesium.

The core of these planetesimals would have taken up something like half of their total volume, and if one sliced such a planetesimal clean in half and looked at the freshly cut face, the core would appear as a circle enclosed within a thick shell of stone. This is the same process by which the Earth's core formed; Mercury, Venus, Mars, and the Moon also have iron–nickel cores. Molten planetesimals, despite being hundreds or thousands of times smaller, were not so different from planets in their internal structures. The spheres of metal and molten stone, big and small, silently orbited the Sun together.

After quickly losing their internal heat to space, the molten rock crystallised and the differentiated planetesimals solidified, forming onion-skin-like bodies with distinct layers. The iron–nickel cores in the centres of the miniature worlds crystallised to form a solid assemblage of metallic minerals rarely found on the surface of the Earth, dotted here and there with penny-sized sulphur-rich blobs. The thick shell of stony mantle armouring the metallic cores also cooled and crystallised, and forged an array of stony minerals familiar to us here on Earth. A thin skin of rock surrounding the outermost layer of the stony mantle capped the surfaces of the planetesimals to form a stony crust. An orderly sequence was formed from inside to out: metallic core, armoured by a stony mantle, capped by a stony crust.

Even once they had cooled, the stories of these planetesimals were far from over. Many shattered during giant impacts and collisions, showering the nascent Solar System with fragments of rock. Some had their stony mantles ripped away during glancing collisions, leaving their metallic cores exposed. Today, more than four and a half billion years later, the Asteroid Belt is still emitting these once molten stones, and some fall to the Earth as achondrite meteorites.

Metallic rocks

Even today, surrounded by technologies that are nothing short of magical by the standards of a few decades ago, holding an iron meteorite still invokes a sense of wonder: they are surprisingly cold to the touch and, at ~ 2.5 times the weight of a similarly sized Earth rock, are unexpectedly heavy. Even now, the weightiness of meteoritic iron still catches me off guard.

French geologist Adolphe Boisse first suggested in the middle of the nineteenth century that iron meteorites resemble the deep interior of the Earth, and he was correct: iron meteorites are pieces of the shattered cores of differentiated planetesimals. That they exist at all reveals a story of calamity. Either the stony mantle encasing the core must have been completely removed by continuous impact and ejection into space, or the entire planetesimal was eviscerated by cataclysmic impact with another body, releasing the naked pieces of iron.

On the surface, an iron meteorite looks like a typical celestial stone – a rock coated in a black varnish-like fusion crust – but sliced open it reveals a shimmering surface of silvery iron–nickel, dotted here and there with small, yellowish, sulphurous blebs, which when polished transforms into the most perfect of mirrors. There is no need to measure their exotic isotopic composition or unpick their strange geological character to see that iron meteorites are alien.

The two major minerals from which iron meteorites are composed – kamacite and taenite – are crystalline mixtures of iron and nickel in different proportions. They are an identical glinting shade of silver, but the beauty of the inter-growth patterns formed by the two minerals is revealed by a specialised chemical procedure called 'etching'. When a chemical mixture of concentrated acid and alcohol is brushed lightly onto the polished face of an iron meteorite, the kamacite and taenite are attacked at different rates. Each tarnishes to a new shade of silver, and an exquisite crosshatch pattern of interlocking metallic minerals is revealed that cosmochemists named a 'Widmanstätten pattern'.[1] Hand-sized needles of kamacite carve through domains of taenite. This unique geological phenomenon is only found in iron meteorites. Some crystals of meteoritic taenite have been known to reach one metre in size, reflecting the incredibly slow rate at which these pieces of planetesimal core cooled over many millions of years.

During my PhD, I was fortunate enough to etch the polished surface of a melon-sized piece of the Campo del Cielo iron meteorite. The ancient Widmanstätten pattern appeared like the image on a developing Polaroid. The revelation of a ~ four-and-a-half-billion-year-old metallic crystal structure is a sight I will never forget.

Another curious property of iron meteorites is magnetism. Molten metallic cores produce powerful magnetic fields, and these fields can extend thousands, or hundreds of thousands, of kilometres upwards and outwards into space. When the cores cooled and crystallised they stopped producing strong magnetic fields, but traces of magnetism were imprinted upon the metallic minerals in the core. The remnant magnetism survives to this day, giving the iron meteorites the ability to deflect compasses and stick to hand magnets.

An unreachable core

Knowledge about Earth's deepest interior largely comes from untangling its complex magnetic field, gravitational characteristics, and the echoes of earthquakes. The Earth has a strong magnetic field, which is due to its partially liquid metallic core; Earth is far too heavy and its gravitational field far too strong for it to be made entirely from the stuff of stone, too. Earth therefore must have a region of extreme density, most easily explained by the presence of a hefty metallic core. Earthquakes provide particularly powerful insights into the Earth's interior; earthquakes at the surface send seismic waves deep into the planet, and by studying how these rumbles echo and ricochet from the different geological layers of the planet's interior, we have built up a detailed picture of the physical characteristics of the mantle and the core.

But while magnetism, gravitation, and seismicity are powerful tools for probing the deepest regions of our home planet, they fall short of getting our hands on a piece of the Earth's core and teasing it apart in minute detail in the laboratory. This would truly revolutionise the way in which we understand our planet. The core of our home planet resides almost 3,000 kilometres beneath our feet, though, rendering it unreachable. If we ever plumb such regions (I do not think we will, but I hope I am one day proven wrong) it will be in the far future.

Until that day arrives, the iron meteorites are the only way we can get our hands on the central core of a planetary body. By teasing apart the shattered pieces of planetesimal cores (like the Campo del Cielo meteorite) in the laboratory, we have come to a deeper understanding of the character of our home planet's core.

The missing mantles

We live on the thin outermost shell of crust that encases the Earth. Beneath this wafer-thin geological layer lies the stony mantle, which extends all the way down to the metallic core at the planet's centre.

The Earth's mantle takes up a staggering eighty-five per cent of the Earth's total volume. The mantle is comprised chiefly of the lithophile elements (like magnesium, silicon, and oxygen) that were left behind as iron sank downwards to the core when the Earth differentiated. These elements predominantly arrange themselves into crystals of olivine and a mineral named ortho-pyroxene, and at great depth rearrange themselves into exotic minerals like wadsleyite, ringwoodite, and bridgmanite.

It follows that once molten planetesimals, like the Earth, differentiated to form metallic cores with complementary stony mantles. It is puzzling and somewhat unsettling that of the ~ 60,000 meteorites currently known to science, none matches the geological make-up of our mantle rocks. Not a single one is comprised of the distinctive interlocking crystals of bottle-green olivine and black-green orthopyroxene. The problem of the missing mantle meteorites is of concern to many cosmochemists. Where are they?

By their absence, the missing meteorites from olivine–orthopyroxene mantles tell a story of abrupt demise. The population of planetesimals that formed mantles must have been obliterated soon after they formed. Their tenure in the Solar

System was brief. Not long after they cooled, almost all of them must have been destroyed by hit-and-run impacts, which either broke them up catastrophically with direct hits, or stripped their mantles away from their cores in glancing collisions. None survive today to fall as meteorites.

Iron meteorites shed some light on the situation. Some rare iron meteorites cooled so quickly – as indicated by the unusually small crystals of kamacite and taenite in their crosshatch Widmanstätten patterns – that they could not have formed in a core insulated by a thick overlying mantle. Their parent planetesimals must have been stripped of their mantles while their cores were still liquid, and, upon losing their insulating rocky shroud, the cores of these planetesimals swiftly cooled to forge undersized metallic minerals.

Unlike the iron shrapnel liberated from their cores, the stony rocks torn from their shattered mantles are physically weak. Upon being ejected from their parental asteroid by an impact, they were abraded by microscopic grains of interstellar rock and quickly dilapidated into crumbly ruin. They were destroyed long ago but their metallic counterparts survive and continue to rain upon the Earth as iron meteorites. Some of the exposed cores of those stripped planetesimals can be found today in the Asteroid Belt,* and from them, newly liberated pieces (such as the Campo del Cielo meteorite) make their journey across interplanetary space to the Earth.

By grouping the ~ 1,200 known iron meteorites together based on similar chemical, isotopic, and geological characteristics, we

---

* At ~ 200 kilometres across, the largest metallic asteroid is asteroid Psyche. In the year 2022, NASA's *Psyche* space mission will depart from Earth on a journey to explore this bizarre world. It is due to arrive in the late 2020s, and it will be the first time humanity has ever explored a metallic world. (I cannot wait.)

know that they come from at least three dozen distinct differentiated asteroids. That is three dozen worlds, each with their own story and geological evolution, that suffered a calamitous end.

The Pallas Iron

From the destruction of the first molten worlds came a brand-new type of asteroid, forged from an intricate jumble of both stone and metal. These are the 'stony-iron meteorites', and the most magnificent among them are the 'pallasites'. This exceptionally rare type of meteorite (only ~ 100 are currently known) is named in honour of the German naturalist Peter Simon Pallas.

In 1772, Pallas happened upon a strange metallic boulder while on an expedition in the Krasnojarsk region of Siberia. A local blacksmith had found the boulder high upon a mountainside twenty-three years earlier, and, believing it would make high-quality smithing material, hauled it back to his village over thirty kilometres away – an impressive feat given that the boulder weighed as much as two grand pianos. But his efforts were (fortunately for the future of meteorite science) in vain: the metal was not workable and was riddled with defects.

Locals told Pallas that the rock was a sacred gift from Heaven, but Pallas, being a scientist, was sceptical. He noted that the boulder was composed of metal dotted with strange yellowy-green crystals and had the mottled texture of a sea sponge. Enchanted by the enigmatic boulder, five years later he arranged for it to be moved to the Imperial Academy in St Petersburg for further study. While there, it became known as the Pallas Iron.

The full title of Ernst Chladni's book *Ironmasses* is *On the Origin of the Mass of Iron found by Pallas and of other similar Ironmasses, and on a Few Natural Phenomena Connected Therewith*. Chladni strongly believed Pallas's stone to be from outer space. A few years after Chladni published his work, a piece of the Pallas Iron made its way to Edward Howard's chemistry bench in London, and alongside the collection of other rocks that allegedly fell from the sky (including the Wold Cottage meteorite), he proved its extraterrestrial origin. The role that the Pallas Iron had to play in the inception of meteorite science was critical.

Composed of penny-sized bottle-green olivine crystals set in a sea of metallic iron, pallasites are the most alluring of all rocks. When a pallasite is sliced into a thin wafer and illuminated from behind, the iron prevents any light from passing through while the olivine transmits a radiant green glow; these slices of stone mixed with metal resemble stained-glass windows ablaze in sunlight. Pallasites look unlike anything normally found on the Earth, geological or otherwise. Their beauty is unrivalled.

For a long while, their origin – exactly where and how they formed on their parent asteroids – was a mystery. Masked by their splendour, the story pallasites reveal is actually one of near catastrophe.

There are cracks marring some of the otherwise pristine olivine crystals. While many are gem-quality minerals – indeed, apricot-sized olivine crystals set in tiaras and necklaces were once popular among the Austrian aristocracy – some are afflicted by deep clefts and defects.

From time to time, a minute droplet of metal is found inside a crystal of olivine. These tiny blebs, like tiny magnets, record ancient magnetic fields, induced while their parent planetesimal was still producing strong magnetism from its molten core. For a magnetic field to be frozen into miniature magnets of this nature it must have been at below a temperature of around 350 °C; any hotter, and the powerful magnetic field emanating from the molten core of the planetesimal would have simply passed through the tiny blobs of metal without leaving a trace.

This observation – miniature magnets trapped inside crystals of pallasitic olivine – has a far-reaching implication. The giant olivine crystals (and the blobs of metal within) must have been cold while the core of their parent planetesimal was still molten, and so must have been situated at shallow depths, far away from the core and towards the surface of their parent asteroid.

So where, then, did the metal that laces the pallasites come from? Metallic iron in such quantities is only found in the deep core – how could it have reached such shallow depths? A simple explanation is that pallasites were forged during colossal impacts between two planetesimals. When the two worlds collided, the still molten metallic core of one – perhaps already stripped of its own mantle by a previous collision – was injected into the olivine mantle of the other. Wrapping around the huge crystals of mantle olivine like red-hot fingers, the metal eventually cooled and solidified, leaving the olivine forever encased in their metallic bind. Rather than being obliterated, the two asteroids mixed to form a hybrid world.

Oxygen isotopes take the story further. The blend of oxygen isotopes within the olivine grains varies from pallasite to pallasite, indicating that the pallasites originate from more than one asteroid. The dramatic mixture of ancient worlds was a process that repeated itself as the Solar System was forming.

This is cause for a moment of reflection on our good fortune: the Earth was once a planetesimal – a small seed from which a planet-sized ball of rock grew – but it survived the onslaught of world-ending impactors through pure chance. Of all the chance encounters that led to the present in which we find ourselves, the survival of the planetesimals from which Earth grew was one of the first. Our existence – the Earth and all life that inhabits it – has been precarious since the beginning.

Thin skins

A few planetesimals remained largely intact throughout the Solar System's violent emergence and went on to become asteroids that survive to this day. For four and a half billion years, they have retained their original onion-like structure – metallic core, stony mantle, and stony crust – and, being unbroken, still send forth meteorites from their rocky exteriors. Upon impact, rocks of a stony nature ('stony achondrites') are ejected from their surfaces into interplanetary space. Originating from the outermost igneous crust of once molten asteroids, these meteorites bear a close resemblance to some types of volcanic rocks on Earth's surface.

It was approaching midnight on 2 September 2015 when one such stone fell out of the sky above the Bingöl province in eastern Turkey. As it plummeted through the Earth's atmosphere, it momentarily lit up the sky, so brightly that it was captured on security cameras 150 kilometres away. Disintegrating in a series of loud explosions, a hail of stones rained upon the village of Sariçiçek. The discovery of crusted black stones scattered in the street and on the roofs of buildings quickly confirmed exactly what had taken place the night before: meteorites had made landfall.

Over the following weeks and months, over fifteen kilograms of rock were recovered of what became known as the Sariçiçek meteorite. It was quickly identified by cosmochemists as belonging to a group of meteorites that are named after Edward Howard: the 'howardites'. There are just over 350 known howardites, and beneath their fusion-crusted surfaces, compared to their achondrite cousins the irons and the stony irons, they look quite – dare I say – unremarkable. Composed entirely of pale-grey granules dotted here and there with small flecks of black and tiny shards of white, a howardite would not look out of place on the floor of a building site. In many ways they resemble pieces of concrete. Aside from the black-varnish fusion crust, unless a howardite had been seen falling from the sky or discovered in a desert, its celestial origin would not be at all apparent.

However, these meteorites are examples of how all rocks – even the most lacklustre – have a story to tell.

When studied as a thin section under a geological microscope, howardites are transformed from pallid grey stones into a kaleidoscopic mosaic of bright and vibrant colours.[2] A chaotic crystalline mosaic of vivid oranges mingled with rich reds, lucid blues and grey-whites, set against an opaque black framework, greets the eye of the lucky observer.

The tiny crystals from which a howardite is composed are an untidy disarray of angular, shard-like fragments with the occasional patch of glass. Most of the mineral fragments are microscopic, and even the largest barely reach one centimetre or so in size. The shards of crystals are a disorganised jumble, in clear contrast to the orderliness of the Widmanstätten pattern in the iron meteorites and the gem-like olivine in the pallasites. Howardites look as though they have been through the mill. The glass confirms that at some point in their geological history, they have been subject to immense shock heating.

The fragmented assemblage of crystals reveals minerals that are exceptionally common on the Earth's surface, like pyroxene, plagioclase, and orthopyroxene. Even if you have never paid any close attention to a rock in your life, you have almost certainly seen pyroxene and feldspar (and probably orthopyroxene) many times. And all three crystallised from once molten rock. Pyroxene and plagioclase are the major minerals comprising basalt – the black-grey igneous rock that forms from frozen volcanic magma – which is found in huge quantities on volcanic islands like Hawaii and Iceland. Orthopyroxene is forged in numerous ways but is common deep inside underground magma chambers, forming as dense crystals that settle out of their parental magma and accumulate in vast piles on the chamber floor.

The minerals in the howardites must have formed in a similar way. But how did the minerals come to be so fragmented? Molten rock normally cools and crystallises into a neatly inter-locking tessellation of crystals. Here on Earth, the original intermeshing texture of an igneous rock is erased and overwrit-ten through weathering and metamorphism, but on asteroids, there are no such processes taking place. There is no wind, rain, or frost to grind igneous rocks into sand, and, being cold, there are no plate tectonics to metamorphose them into new rock types. So how did the howardites become so jumbled? Clues lie in some of the other stony achondrites that fall from the sky.

A meteoritic trio

In 1808, six years after Howard published the first systematic chemical analysis of a suite of meteorites, a fireball tumbled from the skies over the village of Stannern in Czechoslovakia. A flurry of stones showered the ground and many were quickly recovered by witnesses: sixty-six stones, totalling an impressive fifty-two kilograms, quickly made their way into the hands of museums and institutes of science. Cosmochemistry was a brand-new scientific discipline in 1808, and meteorites were highly sought after.

The Stannern meteorite was, at the time of its fall, unique. Beneath the varnished fusion crust lay an ash-grey, fine-grained stone that was soft enough to crumble with bare fingers. Whilst it was stony, it was entirely dissimilar in its geological character from other stone-like meteorites such as that of the Wold Cottage. Scholars quickly noted that it bore a striking resem-blance to a common type of rock found on volcanic islands.

This strange meteorite was made mostly of pyroxene and plagioclase, the principal components of volcanic basalts here on Earth. Stannern was, literally, a piece of basaltic rock that had fallen from outer space. By 1900, a further nine meteorites of a similar geological nature had been spotted falling from the sky across Europe, North America, and India. Being easily discernible from the other (more common) types of stony meteorites, they were named 'eucrites' after the Greek word εὐκρινής (*eucrines*) meaning 'easily distinguished'.

While the minerals from which eucrites are composed must have – like basalts here on Earth – formed in a volcanic system, a casual glance in thin section reveal them to be a jumbled shambles. Absent are the interlocking crystals usually found in igneous rocks here on Earth. The geometric crystal shapes exhibited by pyroxene and feldspar in Earthly basalts are almost non-existent in the eucrites: instead, the crystals are expressed as shattered pieces within the rock, butting up against one another in a random fashion. Many are laced with glass, indicating that at some point in their history, the eucrites suffered great trauma.

In the small hoard of meteorites discovered on the East Antarctic Ice Sheet by Japanese explorers in 1969, one was another unusual kind of stony meteorite. Concealed by the blackened crust of this special find lay an assortment of enormous orthopyroxene crystals, some as big as a good-sized chicken egg, punctuated here and there with the occasional smaller crystal of olivine and feldspar. The orthopyroxene crystals, like those on Earth, must have formed deep underground in a magma chamber. Most of the crystals, while larger than those found in the howardites and eucrites, were a similar angular and chaotic assortment of fragments. There are now almost 500 celestial stones of the same type, and together they form a group of meteorites named the 'diogenites'.

The geological make-up of the basaltic fragments of pyroxene and feldspar in the eucrites are identical to the basaltic fragments found in the howardites. Likewise, the fragments of orthopyroxene comprising the diogenites are identical to the orthopyroxene fragments found in the howardites, right down to the details of their chemical composition. It is as though the howardites are an intimate mixture of two different types of meteorite – the eucrites and the diogenites. If you put a eucrite and a diogenite in a blender, you would make a howardite. This observation can be explained by a simple, yet startling, hypothesis: the howardites, eucrites, and diogenites all originate from the same asteroid.

The three different groups of meteorite have an identical blend of oxygen isotopes, proving (beyond reasonable doubt) that they originate from the same parent asteroid. Together, the **h**owardites, the **e**ucrites, and the **d**iogenites form the 'HED clan' of meteorites, and they permit an unparalleled insight into the geological evolution of their parent asteroid. There are currently over 2,200 HED meteorites in the worldwide collection,[3] and together they have a mass of more than one and a half tonnes. We have four times as much material from the HED parent asteroid as we do from the Moon (*Apollo* samples, *Luna* samples, and meteorites combined). Unlike the irons and the stony irons, the HEDs stem from shallow depths on their parent asteroid, originating no deeper than the outer few tens of kilometres. They are all crustal rocks from the outermost stony skin.

A world of roiling magma

The eucrites tell a story of a blazing hot past and paint a picture of a fiery world. The basaltic nature of the crystals contains clear evidence that – astonishingly – their parent asteroid featured volcanic activity early on in its history. Some of the frozen rocky asteroids we see in the Asteroid Belt today were once world-sized volcanic provinces, whose surfaces would have resembled something like the churning volcanic islands of Hawaii. Fountains of fire erupted from their molten exteriors and sprayed lava high above their surfaces, and as it rained back down to the surface, it collected in large glowing pools. Carpets of lava crept across their surfaces and chilled against the frigid coldness of outer space, cooling to form the fine-grained basalts. Today we recognise pieces of this crust as the eucrites.

Beneath the molten surface of the HED parent asteroid lay the realm of the diogenites. Vast underground chambers of magma churned and roiled, eventually cooling to the point at which orthopyroxene crystals froze out of the liquid rock. Insulated by kilometres of overlying rock and magma, they cooled slowly, growing to enormous sizes before settling downwards into vast piles of large crystals at the base of the underground caverns.

The common feature of all three groups of meteorite is their wreckage: their once beautifully shaped igneous crystals that now exist only as shattered pieces. Like the Moon, the asteroids have no protective atmosphere to shield them from incoming impactors from space, and so, for many billions of years, their rocky surfaces have been pummelled and smashed. This is how the HEDs acquired their splintered geological make-up: impacts breaking up their crystals over and over and over again. Destruction is etched into the geological make-up of the early Solar System.

Impacts on a colossal scale excavated craters deep into the crust of the HED parent asteroid. Fragments of the deep-crustal diogenites were disentombed and brought to the surface, and they mixed with fragments of shallow-crustal eucrites. Together the fragments went on to create the third and final member of the HEDs, the howardites. Over and over again the HED asteroid had its surface punctured, and each time the rocky material making up its crust was turned over and fragmented once more. Microscopic fragments were ground from the once whole rocks, and recombined to make new ones.

The enormous pressure waves that rippled through the fragments during impact events momentarily raised their temperatures so high that tiny amounts of rock melted, instantly quenching to form regions of glass that acted like glue to cement the grains together. The lone and disorganised fragments were recycled to forge whole rocks once more. Many howardites contain distinct fragments that are themselves made from pieces of eucrites and diogenites forged together: they are howardites within howardites, reflecting four and a half billion years of continued and relentless reworking of the rocky surface of their parent asteroid.

Matching meteorites with asteroids

There are more than 60,000 known meteorites and almost 800,000 known asteroids. From the meteorites, we have gained insights into the asteroids from which they originate. But, perhaps surprisingly, we have a poor understanding of exactly which asteroids we have pieces of here on Earth. It is clear from the rich geological diversity expressed in the meteorites that we have samples of many distinct and unique asteroids. Tying a meteorite on the ground to one of the hundreds of thousands of asteroids in space is exceedingly difficult. The HEDs are a rare exception: they are the only group of meteorites that have a strong link with a particular asteroid.

In the 1960s and 1970s there was a concerted effort systematically to characterise the nature of sunlight reflected from the surfaces of asteroids. Different rock types reflect light in characteristic ways, and so by measuring the precise spectral properties of the light reflected from the rocky surface of a celestial body, inferences can be made about its geological make-up. This is geology over interplanetary distances. Many hundreds of asteroids have had their rocky surfaces geologically characterised, but early on, one asteroid in particular stood out as being especially interesting: asteroid Vesta.

As the second largest asteroid in the Asteroid Belt and the fourth to be discovered, Vesta is so big that it can easily be seen through a pair of binoculars, and is often considered by cosmochemists and planetary geologists to be more than a mere planetesimal; it is a planetary embryo. Once its growth into a fully fledged planet was stunted by gravitational interaction with Jupiter, Vesta was locked in its embryonic state, but not before it grew hot enough to melt and differentiate. Despite relentless pummelling by impactors big and small – a few of which, like the Rheasilvia impact, almost fragmented the asteroid into an utter ruin – Vesta survives to this day, making it by far the largest differentiated asteroid, one of the last of its kind.

In the 1970s the rocks of the Vestan surface were found to have a reflectance spectrum characteristic of basalt, indicating that it had an igneous crust made of basalt. Just like eucrites. Portions of the spectrum hinted towards a smattering of orthopyroxene, too, which must have been excavated via impacts from deep into Vesta's subsurface. Just like diogenites. Immediately, speculation that Vesta might be the source of the HED meteorites tore through the scientific community, but Vesta does not lie close to a Kirkwood Gap, and so its location in the Asteroid Belt is not favourable if stones are to be gravitationally delivered to the Earth.

Almost 200 years after Vesta's discovery and around twenty years since the discovery of its HED-like surface, planetary scientists discovered a further dozen asteroids that have remarkably similar reflectance spectra to basalt. One of them was the 4,147th asteroid to be discovered and carries the name 'Lennon' in honour of John Lennon. *Across the Universe*, as it were. These mountain-sized lumps of rock (the largest is around ten kilometres across, fifty times smaller than Vesta) are gravitationally linked to Vesta by their similar orbital characteristics. They were almost certainly ejected from the surface of Vesta by a calamitous impact or two. These small chips off the old Vestan block have since become known as the 'Vestoids', and they can be found orbiting the Sun at between 2.3 and 2.5 au.

None are found orbiting at 2.5 au because here lies one of Kirkwood's 'great chasms'. Any rocky object that finds itself in a Kirkwood Gap – from a mountain-sized Vestoid to a fist-sized piece of Vestan rubble – will swiftly be placed on an orbit taking it into the inner Solar System and on potential collision with Earth. This provides an orbital highway that the HEDs can ride to Earth.

Incredibly, we have had a chance to see Vesta up close in astonishing detail. Between July 2011 and September 2012, NASA's *Dawn* spacecraft explored the surface of Vesta from orbit, circling above the asteroid's fragmented basaltic surface. During thirteen months of close-up investigation, *Dawn* detected eucrite-like and diogenite-like rocks distributed chaotically across the rubbly surface. This strengthened the link between Vesta and the HED meteorites; the latter almost certainly come from the former. The two enormous hollows that afflict Vesta's south pole, the Rheasilvia and the Veneneia impact craters, are likely to be the source of the Vestoids, making these craters the potential homelands of the HED meteorites.

I wish Herschel and his contemporaries were still around so they could see the star-like points of light, that they saw only the other end of their telescopes, become places with their own unique geological evolutions and stories from aeons gone by.

Even more, I wish they could hold meteorites in their hands, full in the knowledge that they were holding pieces of the worlds they were spying from afar.

From exposed cores to fragments of crust and smatterings of shattered mantle in between, the three varieties of achondrites – the irons, the stony irons, and the stony achondrites – have revealed long-lost worlds of metal and molten rock. Far from being lifeless pieces of rocky debris left over from the formation of the Solar System, asteroids are worlds with their own (admittedly short) geological histories. Meteorites give us a glimpse into the destructive forces at work during the formation of our Solar System. Smaller impacts upon the surfaces of asteroids transformed their rocky exteriors into mottled landscapes peppered with craters on all scales, circles on circles, mixing rocks of different types and forging new types of stone. The achondrites exemplify Nature's ability to create new rocks – new geological forms of beauty like the HEDs and the pallasites – from the disassembled pieces.

But tales from the collapsing nebula that were written in the rocky nebula dust from which the asteroids, comets, and planets coalesced, were overwritten when the achondrite asteroids melted. To peer back and discover the stories from the nebula itself – the planetary nursery where the first worlds in our Solar System came into existence – we need to look at meteorites that originate from asteroids that were never subject to the destructive power wielded by heat.

5

COSMIC SEDIMENTS

Sedimentary rocks are wonderful storytellers. For one, they are the custodians of fossilised remains of ancient lifeforms that have graced the Earth over the aeons of geological time. They also contain the annals of environmental change across the planet as landmasses drifted around the globe atop the underlying sea of slowly churning mantle. Sequential layers of rock relay the evaporation of whole seas into expansive planes of salt flats and their eventual replenishment with new waters; they record the torrid heat of ancient deserts and their descent into glacial landscapes; they preserve the ripples from expansive river systems and the shorelines of long lost seas into which they flowed.

Accumulations of loose sediment – whether car-sized boulders or microscopic flecks of silt – are compacted by pressure and cemented by the growth of new minerals to form solid rock. Each constituent grain in a sedimentary rock is older than the rock itself: the grains, by necessity, must have existed before they came to be part of their host rock.

The dust that formed in the collapsing nebula was swiftly destroyed inside the asteroids that melted. Some asteroids, though, escaped melting altogether, and so preserved the dust from which they initially coalesced. It was a playoff between time and size: if an asteroid formed too early, the pinch of short-lived radioisotopes bound within quickly melted it; too big, and its enormous bulk prevented heat from efficiently escaping, warming it to the point of liquidation.

Meteorites that come from unmolten asteroids are called chondrites. Being assembled from countless individual motes of cosmic dust, the chondrites are often considered to be sedimentary rocks, but rather than being laid down in layers in a wind-swept desert or on the ocean floor, they were brought together by the influence of gravity in the protoplanetary disc and compacted to form asteroids.

Chondrites account for something like eight out of every ten meteorites that land on the Earth's surface today. Many of the celestial stones that sparked the history and inception of cosmochemistry were chondrites: Topham's meteorite of the Wold Cottage; four of the eight meteorites chemically interrogated by Howard in his 1802 experiments; eight of the nine meteorites discovered by chance on the East Antarctic Ice Sheet by the 1969 Japanese expedition; and a staggering ~ 85 per cent of all meteorites found in Antarctica since then are chondrites, too. That brings their total number to well over forty thousand.

And while they share a common unmolten heritage, the chondrites are incredibly varied rocks, made up of different combinations of dust types. They fall into three geologically distinct classes: carbonaceous, ordinary, and enstatite. They themselves are further divided into over a dozen individual groups, reflecting the great diversity of the rocky debris in the Asteroid Belt.

Allende

In the small hours of 8 February 1969, a fireball, blazing bright blue with the brilliance of a midday Sun, burst across the sky and illuminated the dark landscape of north-western Mexico.

Eyewitnesses as far away as New Mexico, Texas, and Arizona in the United States saw the spectacular event unfold too, as the fireball travelled northbound at hypersonic speeds. Thousands of stones, each glowing incandescent, rained from the sky as the enormous fireball tore itself into scores of smaller pieces over the Mexican state of Chihuahua. The shockwaves from the detonation rippled through the atmosphere. Some locals, fearing the worst, took refuge in their local churches. The Sun rose to reveal black-crusted stones on the ground, one of which had narrowly missed the post office of the town from which the meteorite would take its name: Pueblito de Allende.*

Within two days, news of the Allende meteorite fall had reached the airwaves in Texas. In an unlikely twist of good fortune, the curator of the Lunar Receiving Laboratory of the Space Center in Houston, Dr Elbert 'Bert' King, heard the reports on his car radio. His laboratory was a brand-new facility at the Space Center (later renamed the Johnson Space Center), built with the sole purpose of housing the rocks that were to be brought back from the surface of the Moon by the *Apollo* astronauts later that year. It was the most sophisticated space-rock laboratory ever built.

King, in a state of hurried excitement, made a few phone calls (with the help of his Spanish-speaking secretary) to newspaper editors in and around Pueblito de Allende, and within hours was aboard a plane to the scene of the fall.

King was determined to recover pieces of the celestial stone as soon as possible, before they had a chance to be subject to the damaging effects of weathering. Fuelled by strong coffee, he arrived on the scene within twenty-four hours of hearing the news, by which point little more than three days had passed

---

* Pronounced 'aye-end-ay'.

since the meteorite fell. What he found was to shape our view of the Solar System for ever.

The first stones seen by King were sitting atop a newspaper editor's desk – one the size of a football – and each was coated in the tell-tale blackened crust. They were undoubtedly meteorites. This meteorite was made from innumerable grains all sandwiched together to form a single rock: it was a chondrite. But it was not just any old chondrite. King recognised it as belonging to an exceptionally rare class of chondrite called the 'carbonaceous chondrites'. Specifically, it belonged to a subgroup of carbonaceous chondrites called the 'CV chondrites'.[*] These rocks are dark-grey and are speckled with the occasional mote of brown and white.

Accompanied by a host of newspaper reporters, who kindly acted as translators, and a policeman acting as a chaperone, King drove to areas where rocks had been found on the ground. Everybody they met seemed to have a piece of the fallen stone and, as the day wore on, the momentousness of the situation became clear. Not only was Allende an exceptionally rare type of meteorite, but the fall was enormous; falls of this magnitude are once-in-a-lifetime events, and to have a cosmochemist on the scene within days was a first.

King left Mexico later the same day with a sack full of stones (almost seven kilograms in total[†]). The precious carbonaceous cargo arrived at the Space Center in Houston a mere 101 hours after it fell from outer space. The arrival of this precious carbonaceous cargo acted as a dress rehearsal for the arrival of the *Apollo 11* Moon rocks that were scheduled to arrive in Houston within

---

[*] The CV chondrites, named after the Vigarano meteorite, are one of the seven groups of carbonaceous chondrites, classified and grouped together on the basis of their geological and chemical characteristics.

[†] A *lot* of meteorite, especially for one as rare as a carbonaceous chondrite.

a matter of months; it also gave NASA scientists plenty of opportunity to test out the brand-new analytical instruments they had prepared for the analysis of *Apollo* samples.

Over the following days, weeks, and months, pieces of the Allende meteorite were collected in their droves by scientists and by collectors from across the globe. Over two tonnes of material was recovered over an area wider than 300 square kilometres, and today it remains one of the largest celestial stones of any kind seen falling from the sky. Thanks to its size and rarity, and the fortunate timing of its fall, Allende is probably the most studied rock in history.

The term 'carbonaceous' is, as it turns out, a historical misnomer: many of the first chondrites seen falling from the sky during the nineteenth century, when meteorite classification was in its infancy, happened to be the carbon-rich varieties. Since then, many non-carbon-rich chondrites have been discovered and classified, but their 'carbonaceous' name stuck. Aside from the occasional oddity, however, all carbonaceous chondrites are sedimentary rocks that beautifully preserve pieces of cosmic sediment that formed in the collapsing nebula. Allende is the most famous of them all.

The word chondrite comes from the Greek word χόνδρος (*chondros*) meaning 'grain'; they are literally 'grain stones'. It is a perfect name, because just by looking at a chondrite it is clear that they are made from innumerable individual grains of dust-sized rock. The largest motes of nebula dust in Allende are typically as big as your thumbnail and they range downwards in size to the microscopic. Individual grains vary wildly in their geological character: some formed in close proximity to the newly formed Sun, whilst others formed in the dark outer reaches of the Solar System in bitter cold. The stories etched inside meteorites like Allende span the entire protoplanetary disc.

First motes

Scattered among the cosmic sediment in Allende are curious snowy-white flecks of rock. They often have erratic shapes. Some are contorted wisps, like flames frozen in time, while others are roughly circular. In many ways, they resemble irregular pieces of fluff afloat amongst the sea of darker grains, but closer inspection reveals them to be an assemblage of unusual minerals rich in calcium and aluminium.

Identical objects had been discovered and described just one year before the fall of Allende in another CV chondrite named Vigarano by French mineralogist Mireille Christophe Michel-Lévy.

Michel-Lévy concluded that these *exceptionnel* snowy-white objects had crystallised via condensation from a red-hot gas at temperatures in excess of ~ 1,400 °C. They were nothing like the minerals one finds in rock that crystallised from liquid magma; these crystals formed directly from a gas. From gas to dust. She had inadvertently stumbled upon the oldest minerals in the Solar System: the first motes of dust to form in the collapsing interstellar cloud from which the planets, asteroids, and comets would eventually form. They have since become known as 'calcium-aluminium-rich inclusions' and are perhaps the strangest and most studied objects found tucked away inside meteorites.

Calcium-aluminium-rich inclusions, or 'CAIs', formed in zero gravity from the cooling cloud of collapsing nebula gas. Crystallising out of the thin wisps of nebula, they would have resembled snowflakes growing out of thin air. In most carbonaceous chondrites, CAIs are tiny white flecks, but in Allende, they are huge, reaching up to several centimetres in length. They are clearly visible, too, set amongst a sea of much darker grains. Much of what we have uncovered about the earliest days of our Solar System has been gleaned from the CAIs in this single meteorite.

Isotopic oddities

In 1973, a team of scientists at the University of Chicago made an astounding discovery. The mix of oxygen isotopes bound inside the crystals in CAIs were unlike anything previously measured in any rock of any sort, terrestrial or extraterrestrial. Testament to the cosmic nature of CAIs, they did not lie along the Earthly terrestrial fractionation line: they were disproportionately enriched in the lightest isotope of oxygen, ^{16}O. In a bucket of ocean water here on Earth, all but around 240 of every 100,000 atoms of oxygen are ^{16}O; in CAIs, that number is 227. The difference may seem small, but by isotopic standards it is huge. Such exotic blends of oxygen isotopes are entirely unknown on Earth. It turns out that almost all pieces of all meteorites have exotic blends of oxygen isotopes, but few are quite as exotic as the CAIs.

It took almost forty years for the full significance of this isotopic discovery to be realised. In 2004, NASA's *Genesis* spacecraft returned to Earth after spending three years in space, during which time it had collected particles streaming outwards from the Sun. After an unplanned crash landing in the Utah desert due to a failing parachute, the samples of pristine solar wind were retrieved from the spacecraft wreckage and, following the most ambitious clean-up operation in the history of science, unspoiled wisps of the Sun's atmosphere were recovered.

The cosmochemists who measured the particles' oxygen isotope composition discovered that the Sun, the star at the heart of our Solar System, contains a blend of oxygen isotopes almost identical to that locked within the minerals comprising CAIs.[2] There is an isotopic kinship between the oxygen in the Sun and the oxygen bound within the CAIs: the CAIs, the tiny white crystalline flecks tucked inside Allende, inherited their oxygen isotopes from the place they formed. They formed next to the Sun. They were almost grazing the rolling and hissing solar surface when they condensed from wisps of nebula.

After condensing out of the glowing gas, these tiny flecks of primordial white Sun snow were blown backwards by fierce stellar winds into the feeding zones of the planetesimals. There, they joined the host of other dusty particles which in time would go on to assemble the rocky asteroids, and some would eventually wind up as the stuff of planets.

Collections of disparate dust

Along with the other grains of cosmic sediment comprising the carbonaceous chondrites, the CAIs are embedded in a cement-like groundmass called 'matrix'. With the aid of a powerful microscope, matrix reveals itself to be an endless collection of nanoscopic dust grains. Most are far smaller than one-thousandth the width of a human hair; but with meteorites, and all rocks for that matter, there is always more than meets the unaided eye.

Co-existing side by side in the same meteorites today, the CAIs and matrix must have mixed as great dust clouds in the protoplanetary disc. Once these billowing masses of cosmic sediment had amassed enough matter, they collapsed under their own gravity to form planetesimals. Despite their close proximity inside chondritic meteorites (they are visibly touching each other), the CAIs and the matrix tell of vastly disparate geological histories. Whilst the CAIs formed beside the scalding-hot Sun, the matrix contains an abundance of material that formed in the freezing cold. This so-called 'volatile' material* formed deep in the dark fringes

---

* In cosmochemistry, a volatile substance is one that easily evaporates into a gaseous state. A good example of a common Earthly volatile substance is acetone, an active ingredient in fumy nail-polish remover.

of the protoplanetary disc where it was cold enough for low-temperature minerals to condense from the nebula gas.

Stellar gusts issuing from the surface of the Sun blew the CAIs far outwards into the colder distal regions of the protoplanetary disc. Rather than streaming sideways along the plane of the protoplanetary disc, the CAIs must have been launched upwards and outwards. Following ballistic trajectories far above the disc, they rained downwards back onto the flat disc plane. Once there, they mixed with the volatile-rich minerals of the matrix to form chaotic and complex dust clouds, which ultimately collapsed to form the jumble of sedimentary planetesimals, too small to melt.

More than 4,000 million years later, some CAIs would escape their parent planetesimal aboard small pieces of rock viciously carved away by an impact, and fragments of the shrapnel would journey back to the inner Solar System. One of these pieces would veer towards the third closest planet to the Sun. After a short but fiery flight through the shell of gas surrounding this ocean world, it would, fortuitously, thump down upon dry land.

The curious creatures who inhabit this world would break open the fallen stones and reveal the CAIs to the light of the Sun once more. They would name this particular stone Allende.

The discovery of deep time

Deep time is unsettling. As animals that have evolved to understand the present on the basis of hours and days, the future on the basis of months and, at most, a few decades, even the concept of what a mere 1,000 years looks or feels like is lost on us. The age of the Solar System, four and a half billion years, is something like sixty million human lifetimes placed end to end; this is a scale where geological timescales bleed into the astronomical.

During our brief residence as a species on Earth we have discovered timescales many millions of times longer than our own short lives. The only way we can precisely ascribe time to geological events – whether it be the evolutionary emergence of complex life, the catastrophic collision between two asteroids, or the deposition of thick ash layers from a violent volcanic eruption – is by dating the rocks that recorded them.

The simple order of a sequence of rocks can take us some way towards dating them by allowing a relative chronology to be constructed. For example, the relationship between layers of sandstone, deposited one on top of the other, can be used to discern the order in which those layers were laid down: the deeper the layer, the older the rock. But this satisfyingly subtle principle can only take us so far down the road of understanding. Relative chronology is akin to knowing that Queen Victoria was born sometime between the birth of Julius Caesar and now; there is no information whatsoever about the actual amount of time that separates the events. The same holds true for the relative chronology of rocks.

To construct an absolute chronology – that is, moving beyond merely putting things in order towards an understanding of how long ago an event took place – we must use the in-built atomic clocks provided by Nature.

Different varieties of atomic clocks are useful for dating different types of rocks over different periods of time, and while they vary in detail, in principle they work roughly the same way. They all use radioactive isotopes, each relying on a radioactive parent that naturally decays to form a non-radioactive daughter. As time inches by and the atomic clocks 'tick', the amount of parent isotope within a rock decreases (because it decays), while the newly formed daughter isotope steadily accumulates. By precisely measuring how much a daughter isotope has accumulated inside a rock, the number of ticks the atomic clock has experienced can be calculated – and then the age of the rock can be calculated. In the case of dating rocks, by convention, we count time backwards from the present day.

Uranium timepieces

To date events that unfolded on a particular timescale, the right radioactive clock must be chosen, in the same way we time events in our own lives using different time-measuring devices. Nobody uses a wall calendar to time the boiling of an egg; similarly, nobody uses a fast-ticking stopwatch to time the passage of months.

Nature has presented us with a radioactive atomic clock that ticks at a rate well suited to dating the events that unfolded billions of years in the past: element number ninety-two, uranium. It ticks (decays) at a rate perfect for dating rocks that formed in our Solar System's formative years.

A given radioactive isotope of any element, including uranium, has a certain probability of decaying in each passing second. This probability is unwavering, remaining at a fixed value, making radioisotopes faithful keepers of time. Upon decaying, the nuclei of both isotopes of uranium – uranium 238 (^{238}U) and uranium 235 (^{235}U) – disintegrate to form new, lighter isotopes of another familiar element: lead, element number eighty-two.

The elegance of uranium's clock lies in its slight complexity. The atomic nucleus of a ^{238}U atom disintegrates to form a particular isotope of lead: ^{206}Pb. Its slightly less massive cousin, ^{235}U, decays to form a different isotope of lead: ^{207}Pb. Both of these new isotopes of lead are not radioactive: they are stable and, once formed, they stick around for ever. Additionally, lead has a stable isotope that, unlike ^{206}Pb and ^{207}Pb, is not produced from radioactive decay: ^{204}Pb. Like its stable cousins, ^{204}Pb sticks around for ever, and once a rock forms the amount of ^{204}Pb within is fixed. The amount of ^{206}Pb and ^{207}Pb in a rock, however, is dynamic, always increasing as the uranium steadily decays.

Both isotopes of uranium decay achingly slowly. If you had a (very) small pile containing 100 atoms of ^{235}U, it would take 700 million years for half of them to decay and turn into ^{207}Pb. A similar pile comprised of ^{238}U would take more than six times longer for half of it to turn into ^{206}Pb: a dizzying four and a half billion years would pass before half of the atoms had decayed. This makes the uranium timepiece perfect for dating events that unfolded over vast timescales.

Thus, the ^{206}Pb and ^{207}Pb in any rock is a mixture of two kinds of lead: the original lead that was there when the rock formed, plus the new lead that has been created from the decaying uranium. If one measures how much extra ^{206}Pb and ^{207}Pb has accumulated in a rock since the rock formed, its age can be calculated. We call the original lead 'primordial' and the in-grown lead 'radiogenic'.

While Earth is important to us, in the context of the Solar System it is just another rock, albeit a large one with a few quirks. Earth, along with the asteroids, contains a mixture of the two types of ^{206}Pb and ^{207}Pb: the primordial lead that was bound within the nebula dust from which it coalesced, and the radiogenic lead that has grown in from the decay of uranium over geological time. If one measured the lead isotopic blend of the Earth today and of the Earth when it formed (the primordial lead), the difference between the two would yield the amount of radiogenic lead which has accumulated over time. It is then possible to calculate how many times the uranium clock has ticked since the Earth formed. From knowing the rate at which uranium transforms into lead – how quickly the clock ticks – the time the clock has been ticking can be calculated. This

would give the age of the Earth. That most human of questions
– *where do we come from?* – would, after 200 millennia of wondering, have a starting point.

The problem with measuring the blend of primordial lead is
that it is no longer available to us here on the Earth. Our planet
incorporated vast amounts of uranium when it formed, and so
the lead we have today is an inseparable mixture of the primordial and the radiogenic lead. The Earth has also continually
roiled and churned over its long geological history, and the
chemical elements and the isotopes of those elements are
constantly being stirred and mixed. Primordial lead has been
mixed with radiogenic lead over and over and over again. Its
signature is lost for ever.

Fortunately for us, Nature provided rocks that have remained
largely unchanged since they formed in the deep past. They are,
of course, the meteorites.

Uranium is a lithophile (rock-loving) element. When the
magmatic asteroids melted and differentiated into their metallic
cores and stony mantles, only the most insignificant quantities of
uranium wound up in their cores. While uranium was expelled
from the metallic cores, appreciable amounts of lead were not.
Having been segregated from one another during the turmoil of
differentiation, the decay of uranium did not affect the amount of
lead – crucially, the ^{206}Pb or ^{207}Pb – in the metallic cores. Therefore,
the lead bound within the iron meteorites, pieces of these cores, was
frozen in when the asteroids formed. It is essentially primordial. By
measuring the blend of lead isotopes in the iron meteorites, the
starting point of the atomic clock can be discerned and the age of
the asteroids can be calculated. This is, by extension, roughly the
age of the Earth and the other planets, too.

It is an incredible thought. By measuring something as seemingly esoteric as the blend of lead isotopes in a metallic rock that
fell from outer space shrouded in flame, the age of the Earth

– the time that story, our story, begins – can be calculated. But as is often the case in science, making the measurements necessary to pinpoint such an important number is an intricate process, fraught with difficulty.

Clair Cameron Patterson

In the year following the end of the Second World War, a nuclear chemist named Harrison Brown was appointed Professor of Chemistry at the University of Chicago. He had spent the previous few years as part of the Manhattan Project, helping develop the technology and scientific understanding needed to create the first instruments of mass destruction: nuclear bombs. After the war, Brown, like many other scientists involved in such a morally charged project, decided to use his expertise to solve some matters of a purely scientific and exploratory nature. The prospect of using uranium – the element which powers nuclear weapons and with which he was well acquainted – to date rocks took his fancy.

Atomic physicists, while making forays into the world of atomic nuclei, had derived an equation that would solve one of the biggest mysteries in science: how old is the Earth? All somebody needed to do was plug in the numbers: the rate at which ^{238}U and ^{235}U disintegrate to form lead; the blend of ^{238}U and ^{235}U on the Earth today; the blend of ^{206}Pb and ^{207}Pb on the Earth today; and crucially, the blend of primordial ^{206}Pb and ^{207}Pb when the Earth formed. The uranium pieces of the puzzle had already been solved, largely during the endeavour to create the nuclear bomb. Modern lead was easy to measure, too, because it is readily available on the Earth. The primordial lead was the final unknown piece of the equation. All somebody needed to do was find, isolate, and measure it.

Brown appointed a PhD student named Clair Cameron Patterson to work on the newly emerging science of precisely dating rocks. Patterson had also been involved in the Manhattan Project and he knew how to operate the instruments used to measure isotopes: mass spectrometers. In an analogous way to how a prism splits sunlight into a spectrum of colours, a mass spectrometer splits rocks into a spectrum of isotopic masses. It never ceases to amaze me that we can precisely distinguish between atoms that differ in mass by only the mass of one single neutron, and then count their relative proportions with razor-sharp precision. Mass spectrometers are technological wonders.

Brown summoned Patterson to his office and told him about the missing piece of the marvellous equation. Brown realised that the primordial lead was hidden away inside iron meteorites: 'We'll get the lead out of the iron meteorite. You measure its isotopic composition and stick it into the equation … and you'll be famous, because you will have measured the age of the Earth.'

Patterson set to work immediately to measure the lead isotopes in an iron meteorite. He spent the next five years perfecting his techniques. Nobody had previously attempted to make measurements like these, and so Patterson, along with the other members of his research group, had to develop brand-new ways of doing things. They had to learn how to precisely measure the blend of isotopes in minute quantities of lead, 1,000 times smaller than anybody had ever measured. They used a common Earth mineral named zircon to practise: zircon is a mineral that is found in many igneous rocks here on Earth, and the ones Patterson practised on were about the size of the head of a pin. If he could precisely measure the isotopic mixture of lead in a microscopic zircon, he knew that he could do it for a precious iron meteorite, too.

But zircon after zircon appeared to contain far more lead than he expected, and Patterson repeatedly gathered nonsense numbers from his measurements. After much frustration and head scratching, he identified the source of the problem to be lead contamination. It was absolutely everywhere. Worryingly, lead was making its way into Patterson's zircon samples during his chemical procedures in the laboratory, and it was swamping and obscuring the indigenous lead that was bound up within his zircon crystals.

A huge effort went into pinning down exactly where this stray lead was coming from, and what Patterson uncovered was frightening: it appeared to be coming from everywhere. There was lead in the acid he used to dissolve his samples; there was lead in the tap water in the laboratory; there was lead in the glassware he used for his experiments; there was even lead in the air, floating around on suspended particles of ordinary dust. It was as though it was coming off the walls. There was lead all over Patterson, too: on his hair, his clothing, his shoes, his skin. And there was lead all over his lab mates.

But Patterson used the problem of lead contamination to his advantage, seeing it as an opportunity to learn how to measure the blend of lead isotopes in many different kinds of materials.

He had just scratched the surface of what would become one of the major environmental issues of modern times: catastrophic industrial lead pollution that affected the entire planet, especially those in urban areas. Leaded petrol, pumped from the exhaust pipes of automobiles into the atmosphere, was a particularly vicious contributor to the problem. So too was leaded paint, which explained why lead appeared to be coming from the walls in Patterson's laboratory. Research into the source and extent of global lead pollution (and the consequential mass poisoning of citizens) was to become the focus of Patterson's work later in his scientific career and was a significant contributing factor to the worldwide phasing out of leaded petrol. The true severity of the lead pollution and its adverse effect on public health was unknown at the time, and so to Patterson for the time being, the biggest problem posed by the ever-present lead contamination was the spoiling of his zircon data. It was all that stood in the way between him and calculating the age of the planet.*

It took five years, but eventually Patterson was successful in measuring the isotopic composition of the lead in pin-head-sized zircon crystals. His numbers finally made sense, but only after ridding his laboratory of lead contamination. He did not trust the cleanliness of the glassware and the chemistry benches in his laboratory, and so he scrubbed them to within an inch of their lives with distilled acids. He did not even trust the cleanliness of the acids he bought to perform his chemical procedures, and so he distilled his own in home-made distilleries. He did not trust the cleanliness of anything.

Patterson went on to build his own brand-new ultra-clean laboratory from scratch to make his measurements of iron meteorites, and went to great lengths to keep it lead free. He even installed a complex filtration system to clean the air of dust particles before it was pumped into the lab, and he enforced a strict dress code: anybody entering his laboratory was forced to wear a super-clean laboratory jumpsuit.

In 1956, after isolating and preparing the primordial lead in iron meteorites in his ultra-clean laboratory, he measured its isotopic composition using his mass spectrometer. The numbers filed out

---

* Next time you fill your car up and see the words *unleaded petrol* emblazoned across the pump, you can thank the work of a cosmochemist in the 1950s. Unleaded petrol is a direct consequence of somebody trying to measure the age of the Earth using meteorites, and is a wonderful example of the good that comes from doing science for the sake of science (so-called *blue sky science*).

of the instrument's chart recorder, and, with bated breath, he plugged them into the marvellous equation. Out popped a number. That number was the age of the Earth, and it was *old*: just over four and a half billion years old. In that moment, Patterson was the first and only person in the history of humanity to know the age of our home planet.

In what has become a classic paper in the fields of Earth and planetary science – 'Age of Meteorites and the Earth'[3] – Patterson published his results and conclusion. The age of the Earth was 4,550,000,000 years, to be exact, and his isotopic measurements were so precise that the margin of error in his calculation was less than two per cent.[4]

The individual pieces of the marvellous equation are worth pausing to consider. Each of them – the rate at which ^{235}U and ^{238}U decay, the development of mass spectrometers, and the means by which isotopic blends can be measured – were born from the development of nuclear weapons. From grotesquely powerful technology capable of extinguishing the flame of human civilisation came a way of plumbing the deep history of our planet and our own origins. In almost destroying our future, we discovered our deep past.

Inception

Patterson and his colleagues dated the Earth's formation, but to date events that unfolded before the Earth formed, one needs a rock that pre-dates even the planets. Fortunately, such rocks survived on the unmolten asteroids: the individual pieces of cosmic sediments that formed in the protoplanetary disc, which are bound up within the chondrites. Being little more than a few millimetres across at most, measuring their isotopic composition is exceedingly difficult, but it can be done.

That isotopes existed at all had only been known for a few decades when Patterson measured the age of the Earth, and so scientists and engineers were still developing the methods by which they could be precisely measured.

Modern clean laboratories and mass spectrometers are far more sophisticated than those available to Patterson in the 1950s. As science marches forward, ever smaller pieces of meteorite can have their ages determined. Samples of lead thousands of times smaller even than those available to Patterson, some mere millionths of one-millionth of a single gram, can now be measured with high precision using mass spectrometers. This has opened up the ability to measure the age of individual motes of nebula dust in chondrites. In this way we can peer backwards to a time before the planets existed, when the first motes of dust were crystallising in the protoplanetary disc.

The trouble with orienting ourselves in Solar System time is pinning down a 'beginning'. As we have discovered, the formation of our Solar System – the story of how the nebula transformed from gas to dust, and then from dust to worlds – was a process rather than an event. It took a few million years. Where do we draw the line between 'before' and 'after' the formation of the Solar System when the boundary between the two is fuzzy? In truth, such lines are arbitrary, largely coming down to a matter of choice rather than being something physically dictated by Nature. Nevertheless, Nature provided an elegant solution to the problem, tucked away among the cosmic sediments that make the chondrites.

The CAIs, the first motes of dust to condense from the nebula, provide us with a benchmark. As the first solids – the first rocks to condense from the swirling nebula – the time of their formation is a natural choice for ascribing a number to the birth of the Solar System and the inception of the rock record.

If one measures the age of the CAIs, one will discover the true age of the Solar System.

Dating a CAI using the uranium clock requires a precise measurement of the blend of lead isotopes within the individual microscopic crystals in an individual CAI. If the lead and uranium isotope composition of each type of mineral in the CAI are measured, the amount of ingrown radiogenic lead can be worked out, and, from that, an age calculated. Analytically, however, this is no mean feat.

CAIs contain only a few billionths of a gram of lead, and so even minute quantities of contamination – so much as one-millionth of one-millionth of one single gram – are catastrophic. Such analyses require laboratories to adhere to the strictest of rules, where everything is aggressively cleaned with distilled acids. Even so, this type of analysis can be done, and the ages of four CAIs have so far been determined. Just four. Dating individual pieces of cosmic sediment using the lead clock is at the cutting edge of modern cosmochemistry.

In 2010, a cosmochemist named Yuri Amelin made the isotopic measurements necessary to calculate the lead age of an Allende CAI for the first time.[5] In his clean laboratory at the Australian National University, he and his colleagues delicately teased apart an individual CAI that was gently plucked from a sliced piece of Allende, and elegantly measured the blend of isotopes bound within. This CAI, named 'SJ101', was (and remains) the oldest rock ever dated using the lead clock. It was over four and a half billion years old: 4,567,180,000 years old, to be precise. This represents the moment at which our twenty-four-hour geological day began.

The margin of error on this measurement – half a million years either side – is as impressive as the age itself. This sounds like a huge uncertainty in absolute terms, but relative to the age of the CAI, it is tiny. In our twenty-four-hour geological day, fifty million years shrink down to a mere fifteen minutes.

Since Amelin and his colleagues determined the lead age of the CAI from Allende, three more have been dated. These CAIs were from a different meteorite – another CV chondrite named Efremovka that was unearthed in Kazakhstan in 1962 – and they yielded the same number: 4.567 billion years – old even by geological standards.

The time at which the CAIs condensed from the protoplanetary disc is, at least for now, used to define the age of the Solar System. Time zero: 4.567 billion years. The inception of the rock record, and the moment our story begins.

Our home is a system of planets orbiting a star. Even the name Solar System is derived from the Latin word *sol* meaning Sun. Home is, in a real sense, a star system. Despite that, we define its age using CAIs, solid crystalline objects that condensed from the cooling cloud. We define the age of our star system using not a star, but rocks.

Almost everybody at some point in their childhood gasped, wide-eyed, upon learning how long ago the Great Pyramids in Egypt were built (~ 4½ thousand years ago), or how long ago the dinosaurs became extinct (~ 64 thousand thousand years ago), or how long ago the Earth formed (~ 4½ thousand thousand thousand years ago). Our discovery of deep time and the age of the Solar System in the meteorites marks one of the many triumphs of science.

Our ancestors erected monumental henges made of stone to predict the motions of the stars into the future. Naturally occurring atomic clocks, ticking away inside rocks that fall from the sky, are how we put a timescale to our Solar System's story backwards into the depths of the past.

There is an epic written inside every piece of meteorite. The oldest of these tales, which begin with the condensation of the Sun-grazing CAIs, are written within the cosmic sediments which clustered to form the asteroids. Chondrites each contain billions of years' worth of Solar System history. This history is strung together on the same unbroken thread of geological time: from the formation of nebula dust to the formation of the Earth, and from the creation of artwork in the Blombos Cave to our present day.

6

DROPS OF FIERY RAIN

Meteorite classification is, like most systems of categorisation, laced with historical hangovers and misnomers. Just as the name for carbonaceous chondrites is misleading in that not all carbonaceous chondrites contain significant amounts of carbon, the most abundant type of meteorite found on the Earth today is also mislabelled. These are the 'ordinary chondrites', so-called because they are by far the most numerous type of stone falling from the sky, making up something like eighty per cent of all known meteorites. But do not be fooled – they are by no stretch of the imagination 'ordinary'.

Some of the most important meteorites – Topham's Wold Cottage meteorite, which helped capture the interest of the scientific community; the Czechoslovakian Příbram meteorite that proved the Asteroid Belt to be the source of most celestial stones; eight of the nine Antarctic meteorites that sparked the Antarctic meteorite gold rush in 1969 – were ordinary chondrites. So too were some of the oldest recorded meteorite falls.

Old falls

The oldest witnessed meteorite fall for which written accounts exist touched down in Japan in 861 AD. Written accounts tell of a peaceful night suddenly disturbed. A flying object, tumbling to the Earth in a brilliant flash of light and a deafening boom, crashed through the roof of a shrine in the village of Nogata-shi. The sacred object was placed in the custody of shrine officials where it remained, inside a wooden box, for more than 1,000 years. Upon cosmochemical analysis and classification as an ordinary chondrite in 1983, it was named 'Nogata' after the city where it fell.

Over 600 years after the fall of Nogata, a fall was recorded at Ensisheim, a city on the French-German border. It was just before noon on 7 November 1492 when the air was filled with a terrible sound. Echoes of a thunderclap reverberated as far away as the Swiss Alps, lying some 150 kilometres to the east. It was unlike anything anyone had ever heard.

The sole witness to what caused the noise was a young boy, who watched wide-eyed as an enormous object plummeted from the sky into a wheat field just outside the city walls. The people of Ensisheim rushed to the scene, and discovered a great thunderstone sitting in the bottom of a pit. The 130-kilogram stone was scarred by traces of fire. When it was hauled out of the ground, the crowd immediately began chipping pieces off. They thought the strange stone held magical properties and would bring them good fortune.

Of course, nobody believed that the stone had actually fallen from the Heavens. All they had to go on was the word of a (clearly doolally) boy. Even 300 years later, six years after the publication of Chladni's *Ironmasses*, French chemist Charles Barthold concluded that the stone of Ensisheim had washed down from the nearby Alps and dismissed the miraculous accounts of the time. He said they were nothing more than the ravings of superstitious townsfolk.

The Chief Magistrate of Ensisheim quickly arrived at the scene of the fall and put a stop to the destruction of the stone by the excited crowd. It was hauled back to the city and placed in the doorway of the parish church, where it remained, bound in iron chains, for three centuries until 1793 when French revolutionaries commandeered the meteorite and temporarily moved it.

The stone currently rests inside a beautiful cabinet of wood and glass in the Musée de la Régence of Ensisheim. Only fifty-four kilograms remain of the original 130. Fortunately, the stone is now protected by a special order: la Confrérie Saint-Georges des Gardiens de la Météorite d'Ensisheim (the Brotherhood of Saint George of the Guardians of the Meteorite of Ensisheim). Dressed in red capes and beige plumed hats, the *Gardiens* care for the meteorite, preserving its splendour and intrigue for future generations. Ensisheim, like Nogata, is an ordinary chondrite. And they, like all 52,000 other known ordinary chondrites, are far from ordinary rocks.

Like the carbonaceous chondrites, the cosmic sediments in ordinary chondrites – the dust from which their parent asteroids coalesced – were never unmade by intense heat. Remaining almost pristine, they too are custodians of the rocky dust from which the asteroids and planets coalesced. They are, however, geologically distinct from the carbonaceous chondrites.

Beneath the thin skin of fusion crust, carbonaceous chondrites are dark grey, almost black. Many closely resemble coal. Ordinary chondrites, on the other hand, vary from pale grey to a beige-brown and, glinting with the occasional fleck of metal, they look like granular pieces of biscuity stone.

Grouped together on the basis of their isotopic, chemical, and geological similarities, the ordinary chondrites fall into three groups with progressively lower amounts of iron: the high-iron 'H' group, the low-iron 'L' group, and (imaginatively) the low-low-iron 'LL' group. These groupings are reflected on a deeper level, too: each hosts a similar, but still clear-cut, blend of oxygen isotopes and each group probably comes from its own distinct parent asteroid. Somewhere out there in the Asteroid Belt is an asteroid from which the H group comes, one from which the L group comes, and one from which the LL group comes.

Upon slicing open the most pristine samples, it is immediately clear they are cosmic sedimentary rocks. They are comprised almost entirely of individual circular globules, which are countless in number and sandwiched together to form solid stone. When plucked whole from a chondrite, the circles reveal themselves to be spherical beads.* They are typically poppy-seed sized but range in magnitude from anywhere between the head of a pin and a large garden pea. They are common in carbonaceous chondrites like Allende, but they entirely dominate the fabric of the ordinaries. They take their name from the meteorites in which they are found – chondrites – and we call them 'chondrules'. Every chondrite, from the carbonaceous to the ordinary (and the others in between), contains chondrules, but they are by far the most numerous in the ordinaries.

A gentleman scientist

Four of the rocks bestowed upon Edward Howard during his systematic cosmochemical analyses of meteorites in 1802 were chondrites. Each was ordinary, although this was long before meteorite classification had named them so. Howard noted in his publication that three contained 'small globular bodies' made from a glass-like substance. About the small globs, he noted: 'they have a perfect smooth and shining surface, so as very often to present the appearance of small balls of glass'.

Nothing of their sort had ever been observed in any rock. Spherical pebbles and grains of sand, usually rounded by the abrasive action and jostling of moving water, are often found in

---

* As an orange sliced down its centre appears circular when viewed face on.

sedimentary rocks here on Earth. But such sediments are not glassy like the ones found inside the chondrites. Besides, there are no flowing rivers or oceans to jostle the rocks of the asteroids into rounded spheres.

Henry Clifton Sorby was born in Sheffield in 1826, almost a quarter of a century after Howard published his seminal work. He grew up in a Europe where scientists had fully embraced the celestial origin of meteorites (though exact details were still being hotly debated). After leaving school aged fifteen, Sorby was home schooled. He never attended university, but he had a keen interest in the natural sciences. He was an only child born into an affluent family, and so after his father's untimely death, Sorby found himself aged twenty-one with a comfortable income and no need to work for a living. Instead of investing his money in business ventures, he dedicated his inheritance to his passion: science. In his 1874 acceptance speech on being awarded the Royal Medal by the Royal Society, Sorby explained: 'When I commenced life as a young man I had the choice of either wisdom or riches, and resolved to content myself with moderation and devote myself to science.'

Sorby purchased some equipment and instruments to furnish a brand-new workshop and built a science laboratory in his family home. He spent the rest of his life working there, right up until eleven days before his death at the ripe old age of eighty-two.

Already well-acquainted with crystallography and the study of minerals, Sorby focused his efforts on the creation of geo-logical thin sections in order to study the fabric of rocks. Carefully experimenting with different ways of grinding slices of stone down to a thickness of just thirty microns – fractions the width of a human hair – Sorby developed rock-preparation procedures that are still in use today.

Sorby's samples were of exceptional quality. In his 1850 paper on the fine sandstones of the Yorkshire coast he demonstrated how deep insights into the geological character of rocks could be gleaned by examining their microscopic structures. From its small-scale geological character, inferences could be made about how a rock came to be. It was Sorby who pioneered the means by which we read rocks on this scale.

A decade after his enormous contribution to Earthly geology, Sorby turned his geological skills from the rocks in the ground to those from outer space, encouraged by the astronomer Robert Phillips Greg, who had recently published a compilation of all known meteorites and fireballs. Initially, Sorby set out to understand the best-known feature of all meteorites: the charred and blackened fusion crust. Despite being a ubiquitous and unique characteristic of meteorites, little was known about this strange and unusual geological feature.

Skin deep

Taking a slice of meteorite and crafting a thin section from it, Sorby examined the fusion crust in more detail than anybody before him. He noticed that the crust was made almost entirely from glass. Sorby knew that it could only mean one thing: blistering heat at some point during a meteorite's lifetime. The most obvious place this could happen is during a meteorite's incandescent flight through the atmosphere, where the exterior of a meteorite must entirely melt before instantly quenching into a glassy skin. The meteorite is travelling at hypersonic velocities, and so it is no wonder the temperature is high enough to melt solid stone.

Earth's atmosphere acts as a brake, slowing a streaking meteorite as it plummets. It is this braking action that can burden a meteorite with enormous strain, causing large ones (like Allende) to be torn to pieces as they decelerate. At ten to thirty kilometres above the Earth's surface, all cosmic velocity is lost, and the meteorite continues its fall to the Earth at a lowly 0.1 kilometres per second – the speed of any stone dropped from a great height. Upon slowing, the meteorite's molten exterior quenches in an instant and attains its glassy form.

Once a meteorite has slowed to the speed of an ordinary falling stone, the incandescence fades and it enters 'dark flight'. Here, it completes the final leg of its journey at modest speeds. Its trajectory is now subject to the whims of the weather, with wind direction playing a huge part in where it ultimately touches down. It is for this reason that even meteors and fireballs caught on camera are difficult to find – their trajectories during those final few minutes of descent are exceptionally difficult to predict when even modest winds can blow them miles off their original course.

The shape of a falling meteorite also affects its path during dark flight in unpredictable ways. Like Earth rocks, meteorites come in all shapes: some smooth and rounded; some lumpy and covered in pits that resemble thumbprints pressed into soft plasticine called 'regmaglypts';* others are cone shaped with furrows radiating from the central peak like the spokes of a bicycle. Many meteorites look as if they have been carved from soft wax.

With a powerful microscope and sophisticated optical equipment at his disposal, Sorby also measured the thickness of the fusion crust. It was comparable to the thickness of an apple's skin, barely one millimetre. The rock's interior, however – even right up to the fusion crust – was completely unaffected by the heat of atmospheric entry. Here and there, angry tongues of

---

* From the Ancient Greek words ῥῆγμα (*rigma*, meaning cleft, or a breach) and γλυπτόν (*glypton*, meaning carving). Regmaglypts form as spiralling vortexes of red-hot air carve depressions into the exterior of a falling stone.

fusion crust glass protruded a short distance into the meteorite, injected between the rocky crystals by the immense pressure, but the rest of the stone was totally unaffected. A meteorite's molten surface is stripped away before the heat has a chance to work its way inside the stone.

Alexander Stewart Herschel, the famous astronomer William Herschel's grandson, chronicled a fine example of an ordinary chondrite that fell from the skies above North Yorkshire on 14 March 1881 in the June edition of the *Monthly Notices of the Royal Astronomical Society*. He described the stone as a 'beautifully perfect meteorite' whose 'thin black molten crust ... hides from the eye its true stony character'. By this point, the tradition of naming meteorite falls after where they landed had caught on, and so this meteorite was named 'Middlesbrough'.

The Middlesbrough meteorite landed mere metres from the tracks of the North Eastern Railway. Masked by the cold light of the bright mid-afternoon sunshine, the fireball accompanying the plummeting stone was invisible, but it made plenty of noise during its descent. Explosions were heard emanating from the skies above North Yorkshire more than thirty kilometres away. Workmen, carrying out maintenance on the railway line, heard a loud rushing sound overhead that culminated in a deep and heavy thud somewhere close by. They discovered the fist-sized stone at the bottom of a hole in the railway embankment. The stone had embedded itself through a forearm's length of earth, and the hole was easily wide enough to swallow a man's arm.

When he saw the stone only days after it fell, Herschel immediately recognised it as an exceptional meteorite, and commented: 'whatever provision is finally made for its preservation and mineralogical examination and description, it should not undergo more defacement from its original integrity than is absolutely necessary'.

The Middlesbrough meteorite, classified as an ordinary ('L') chondrite, really does look alien. Furrows, crafted by the blistering air during its fiery descent through the atmosphere, emanate from a central point in the middle of its exterior. Thick lobes of molten crust encasing the stone resemble tongues of lava creeping across a barren landscape, and flow lines look like the roots of a tree scorched by a forest fire. The stone is bright black.

There ensued a mild spat between several scientific institutions as to where the meteorite should be housed. Ignoring the pleas from the British Museum and Durham University, the board of the North Eastern Railway Company saw the meteorite as belonging to the county where it fell: Yorkshire. A few grams were taken from the stone and given to the British Museum in September 1881 and the main mass was duly presented to the Yorkshire Philosophical Society.

It was placed proudly on public display at the Yorkshire Museum in York for all to see. There it remains to this day. Even to the untrained eye, the Middlesbrough ordinary chondrite is an extraordinary-looking stone.

Fiery showers

Sorby delved deeper. Working inwards from the charred exterior of the fallen stones, using the tools and knowledge he had acquired during his unpicking of Earthly rocks, he set to work, peering down his microscope at thin sections of meteorite in his home-made laboratory. He would have unparalleled insight into the geological nature of the tiny spherical beads from which they are almost entirely made: chondrules.

The power of the geological microscope, equipped with its series of polarising filters and lenses, transformed the rocky beads in Sorby's thin sections of ordinary chondrites into kaleidoscopes of vibrant colour. Chondrules were unlike anything he had ever seen. They were unlike anything anybody had ever seen, because nothing of their like exists in any rock made on Earth. Looking down his microscope, Sorby saw that the circular poppy-seed-sized beads were actually made from a complex assembly of minute crystals. In many ways they resembled igneous rocks. Most were composed of crystals reminiscent of rocks forged from magma here on Earth – but they varied enormously in their geological characters.

Some chondrules consisted of large tombstone-shaped crystals of olivine set in a sea of otherwise featureless glass; others were set in a cross-hatch pattern of needle-like crystals. More often than not, the olivine crystals were small and tightly packed together, leaving little room for intervening glassy material. When viewed through a polarising filter, the olivine interference colours shone like beacons, glowing in luminous hues of purples, blues, and greens.

A few unusual chondrules housed an array of crystalline needles that radiated from a single point on the circular outer edge. It was as though they were once spherules of perfect glass that had been shattered, and now had crack-like needles radiating through their structure. On a rare occasion, a chondrule that had no recognisable crystal structure was spotted, appearing as a pitch-black opaque circle when viewed through a polarising filter. Such chrondules were made entirely of featureless glass with no internal crystals.

My own cosmochemistry research to date has largely focused on chondrules (I have a soft spot for them). The textures Sorby described in the nineteenth century are exactly the same textures that I see in chondrules today; some of the descriptions of my samples are practically identical to Sorby's descriptions of his samples. When I read his words it is as though he is describing the same thin sections I work on in the laboratory, even though our lives are separated by a century.

Sorby concluded that chondrules formed from molten rock: each minute cosmic bead is a microcosm of igneous crystals. Their igneous textures mean that they must have once been liquid, and the glass present in practically every chondrule suggests that the chondrules cooled rapidly from their molten state (but not quickly enough to entirely stop some crystals from growing). Heat played a pivotal part in a chondrule's story.

Sorby also noticed that despite their wildly varying internal textures, each chondrule was more or less circular, independent of its interior geological crystal structures. This simple observation has huge ramifications: the chondrules must have been separate entities when they formed, and they only coalesced to form asteroids after they had frozen to form solid rocky beads. If they were tightly packed together (as they are today) when they were still molten, they would have melded together and pressed into one another to form a patchwork of warped grains. But they were, for all intents and purposes, perfectly round. They were once detached, and not touching one another at all while they were still hot and molten. Sorby shared his newfound insights with fellow scientists in March 1877, commenting that 'the constituent particles of meteorites were originally detached glassy globules, like drops of fiery rain'.[1]

Sorby turned out to be absolutely correct in his conclusion. The chondrules were once individual droplets of molten rock, free-floating and adrift in the protoplanetary disc.

Precursors

The material that was heated and welded together to form chondrules probably existed as miniature clusters of fine dust: imagine vast clouds of tiny cosmic dust-bunnies afloat in the protoplanetary disc.

When an igneous rock (such as a chondrule) cools from its parental molten state, the atoms in the liquid lock together to form new crystals. The newly formed minerals crystallise in a predictable and orderly chemical sequence: one type of mineral grows, followed by another, followed by another – a steady progression. Inside chondrules, however, sometimes there is a mineral completely out of step. These odd minerals have a chemistry that is out of sequence with the rest of the crystals and glass, and so there is no way they could have crystallised inside their host chondrule. They are foreign objects that somehow made their way inside their host crystalline bead.

The only feasible way that these out-of-place minerals could have wound up awkwardly sitting inside a chondrule is if they are pieces of surviving chondrule precursors. They are part of the dust-balls – small pieces of primordial nebula dust – that survived being melted and welded together during chondrule formation. Today they exist as relics of the dust from which chondrules were forged. The flash heating and searing temperatures that subjected the dust-balls to a maelstrom of reworking were not quite enough to melt them entirely, and so these relics survived.

Residual cosmical matter

It turns out chondrules vary in their isotopic composition. It seems that every other element shows some strange and varying blend of isotopes: oxygen, chromium, titanium, sulphur, tungsten, and unfamiliar elements such as molybdenum and barium. The isotopic composition of chondrules are all over the place, which means that they must have formed in different parts of the protoplanetary disc.

Even chondrules that are tucked inside the same meteorite may have formed in vastly different parts of the disc: turbulence, sweeping chondrules this way and that way, probably blew them far from where they initially crystallised. Hundreds of millions of kilometres may have separated their original homelands before they mixed, and today they are found touching each other in the same thin section of meteorite.

Upon cooling and crystallising, the newly formed chondrules became part of the dust clouds, alongside other cosmic sediments like CAIs, that collapsed to form the planetesimals. Many chondrules would eventually go on to build planets. Sorby glimpsed this too: 'meteorites are the residual cosmical matter, not collected into planets'.

The tiny spherical beads of rock from which the chondrites are mostly made are, by extension, the major building blocks of planets. It is a startling discovery of cosmochemistry that a rock the size of the Earth was assembled largely from poppy-seed-sized igneous beads. Some million million billion billion chondrules were collated to build a planet the size of Earth. They were, of course, destroyed by the ruinous power of heat as the Earth formed, but the chondrules that formed the unmolten asteroids, against all odds, survived. We find them today in the chondrites.

If there was no dust (such as CAIs and chondrules) populating the protoplanetary disc, there could be no planets; by definition, the building blocks must pre-date the thing they are building. By necessity, this means that at least some of the cosmic sediments, including the chondrules, must be older than the planets. Therefore, the tiny igneous spherules packed inside ordinary chondrites pre-date the Earth, and so by unpicking their geological character, we plumb the pre-history of our home planet.

Clair Patterson needed large quantities of rock to perform a single isotope analysis when he dated the Earth. Precisely measuring the age of individual poppy-seed-sized pieces of cosmic sediment – such as a CAI or a chondrule – was impossible in the

1950s. But it is possible now. We discovered in our foray into the cosmic sediments that the CAIs are the oldest dated rocks and their age is used to define the age of the Solar System. But what about the chondrules? Where do they fit in?

Of the three dozen chondrules that have been dated using the uranium isotope clock,[2] none pre-dates the CAIs. A few, however, have the same age as CAIs, which means that some chondrules began forming alongside them. But while the CAIs (apparently) all condensed at the same time, the ages of chondrules vary enormously.

The first chondrules to form are as old as the Solar System itself – 4,567 billion years old – and the last ones to form did so some five million years later. Five million years; within this short space of time, practically all of the dust – all of the rocky building blocks of the Solar System – from which the planetesimals and planets coalesced, had been forged, and the chondrule factory had been switched off. In our twenty-four-hour geological day, all the dust formed within the first ninety seconds. Nothing brand new was made after this time; things were only recycled into planetesimals and planets.

Billowing through the protoplanetary disc as a mass of brightly glowing droplets of lava, clouds of freshly sintered chondrules swarmed for five million years. Within days of melting, each chondrule had cooled and solidified into a miniature igneous microcosm of crystals. The incandescent clouds of chondrules dimmed as they cooled. Alongside CAIs and other grains of cosmic sediment, the chondrules coalesced to form dense thickets, and eventually the first rocky worlds. Trillions upon trillions of chondrules, in numbers that far exceed the number of stars in the observable Universe, spiralled as gravitational vortices, and coalesced to build the asteroids and the planets. What a sight it must have been. As the final few residual pieces of nebula dust were swept up by the rocky bodies orbiting the Sun, the final few chondrules pelted downwards upon the newly forged rocky surfaces like drops of fiery rain.

By the time the final few chondrules had formed, the era of the nebula was over, and the Solar System had evolved from a formless cloud of gas into a flat disc able to forge fully fledged planets. The first chapter of our Solar System's history – from gas to dust – was brief.

Many chondrules became part of the planets that grace our night skies. The ground beneath our feet owes its existence, in large part, to poppy-seed-sized spheres of quenched rock.

A long-standing mystery

That the chondrules formed from the welding together of tiny dust-balls in the solar nebula gives rise to the question: what melted the dust-balls in the first place? This is one of the longest-standing and contentious questions in cosmochemistry.

As Sorby studied the character of chondrules in his thin sections of ordinary chondrite, he puzzled as to how they had been created. He shared his bewilderment with his colleagues: 'these [chondrules] do not either by their outline or internal structure furnish any positive information respecting the manner in which they were formed'.

He recognised that they must have once been superheated in some sort of incandescent atmosphere under 'very special physical conditions'. Other than planets which had yet to form, the only place in the Solar System where the temperature necessary to fuse rocky dust particles into molten droplets existed, was, he reasoned, the Sun. By the time Sorby was making his microscopic observations of celestial stones, telescopic observations of the Sun had revealed flares emanating from its surface. Some of those flares were large enough to engulf the Earth. He hypothesised that these solar explosions were enough to detach freshly made chondrules from the Sun's surface and blow them backwards into the realm of the planets.

Remarkably, although he had no way of knowing at the time, Sorby had inadvertently described CAI formation. Today we find that the blends of oxygen isotopes within chondrules are entirely different from that of the Sun, however, and so formation close to the Sun cannot be the case. Chondrules must have formed further away from the young star in the depths of the spinning protoplanetary disc, closer to the regions where the planetesimals were forming.

What could have flash heated nebula dust past its melting point far away from the Sun, in the colder reaches of the disc? How did the dust-balls of chondrule precursors come to be molten? What sort of transient heat source could reduce thickets of dust to molten droplets?

Nobody knows.

Chondrule formation is one of the most wonderfully perplexing unanswered questions about the nature of meteorites. Exactly how small tufts of nebula dust were turned into molten droplets of rock remains a mystery, but there are some good guesses.

We know that in interstellar nebulae where stars are forming, stars are also dying. Massive stars end their short lives in spectacular explosions, and we know that our own Solar System formed in the vicinity of stars that suffered this fate. Perhaps it was the shockwaves from a nearby stellar explosion, rippling through the protoplanetary disc, that flash heated the chondrule dust-balls. If this is true, then the death of a star may have triggered the birth of the planets.

We also know that the protoplanetary disc was incredibly dusty in places. As the minute motes rubbed past one another while orbiting the Sun, they could have accumulated tiny amounts of static electricity (just as rubbing your hair with a balloon will cause it to become charged with static and stand on end). If enough electrical charge accumulated then huge electrical imbalances might have arisen and, in a bid to correct the lopsidedness, enormous electrical discharge events may have occurred. Sheets of lightning, tearing through the dust, may have been sufficient to melt them and form chondrules. The disc would have temporarily illuminated the dark outer regions of the Solar System, and glowing droplets of molten dust would have been left in the lightning's wake.

Or perhaps it was an altogether different process that flash heated the dust. We know beyond reasonable doubt that there were planetesimals orbiting the Sun almost immediately after the Solar System formed: atomic clocks recording the timing of planetesimal melting in iron meteorites place their formation within one million years of CAIs (time zero). Maybe these early formed planetesimals, the first rocky worlds to inhabit the Solar System, flash heated chondrule precursors as they careened through the dust clouds like the prow of a ship ploughing through the ocean.

These are just three of the many models explaining the origin of chondrules. While each idea successfully explains a great swathe of chondrule characteristics, not one successfully explains them all. It is likely that in reality, things are more complicated and the chondrules formed by several different means.

To unpick the story concealed by the chondrules once and for all will require a multi-pronged approach and the scientific mindset drawn from many different disciplines. While it was a nineteenth-century geologist who first posed the question of how the tiny rocky spherules came to be, the traditional tools of geology are probably not enough to find the answer. Cosmochemistry has since taken the story further with sophisticated analytical instruments, but even those can only take us so far. Nowadays, experimentalists are in the game, and they simulate the conditions of chondrule formation in their laboratories, carefully creating imitations of the environments that existed in the protoplanetary disc. Physicists and mathematicians are working on the problem, too, simulating the conditions of the protoplanetary disc inside powerful computers; they can fine-tune and tinker with nebula environments fabricated from computer code to create chondrules in a virtual world.

Chondrules dominate, with rare exception, the rocky fabric of every single chondrite. In that we find them to be the dominant building blocks of so many meteorites – which come from asteroids – by extension they must also be the dominant building blocks of planets. They were pivotal pieces of planet formation. They must have swarmed far and wide as our Solar System was aggregating 4.6 billion years ago; from the torrid heat of the inner Solar System, all the way to the frigid fringes of the outer. They were absolutely everywhere.

But for now, the mystery of chondrule formation represents an enormous gap in our understanding of how planetesimals – and ultimately the planets – formed. We are far from a concrete answer as to how these tiny beads of rock were forged. How the tiny thickets of dust came to be welded by the torment of heat remains a gaping hole in our knowledge of the Solar System's story. Being made at least in part from chondrules (and probably in large part), this is also a gaping hole in the story of how our home, the Earth, came to be. But I am convinced that one day they will give up their secrets.

Throughout the history of science, answers and understanding often come unexpectedly. New paradigms sweep across entire disciplines with little forewarning. Even two centuries ago, who could have guessed that the stones falling from the sky held such secrets? Who could have foreseen that from them we would read such wonderful tales of antiquity? There are certainly many more such secrets and tales that, at least for now, remain hidden from view.

In the early 1960s, almost a century after Sorby first described the spherical crystalline products of flash heating, another new and unexpected discovery was made inside the chondrites. These strange objects, so minute that they were invisible to Sorby and his contemporaries, take us back in time much further than the formation of the chondrules and the CAIs. In fact, they take us much further back in time than the formation of our Solar System itself.

Cosmochemists, by delving deep into the fabric of these celestial stones, discovered a way to get their hands on rocks that formed outside the Solar System entirely.

7

STARS DOWN A MICROSCOPE

Almost all the ninety-odd chemical elements that craft entities in the Universe – from trees, to rocks, to asteroids, to entire systems of planets – owe their existence to naturally occurring nuclear reactions. Every one of these reactions, which take place on the atomic level, happen inside stars in a process we call 'nucleosynthesis'. Stars are the crucibles in which chemistry first came into being.

Stars are a part of our Solar System's story in a way that far exceeds the human-centred narratives of religion and myth. We now learn the deepest foundations of our story not in constellations or from fictitious sky-dwelling deities, but in the light from individual stars down the eyepieces of telescopes. Parts of this story also make themselves known in the stones that fall from the sky.

Simple beginnings

Some 13.9 billion years ago, the Universe came into being during the Big Bang. This marked the beginning of everything. Earth's geological history is a sub-plot of the Solar System's history, which itself is a sub-plot of the story of the Universe.

Within fractions of a second of forming, the nascent Universe was awash with a sea of subatomic particles: protons and neutrons, the constituent parts of atomic nuclei. For the first minute or so, the protons and neutrons endured as solitary particles, and so the only chemical element in existence was element number one – hydrogen.

Initially white hot from the energy of the colossal bang, the protons and neutrons remained separate, ricocheting off one another in an intricate dance of subatomic collisions. But as the Universe expanded and cooled, interactions between protons and neutrons became increasingly tepid, and nuclear forces began knitting them together. Within minutes, they began to combine to create a new form of matter. Helium, element number two, was coaxed into existence.

Within fifteen minutes of the Big Bang, the expanding Universe had cooled by a factor of 100 million million billion, and the nuclear reactions between protons and neutrons ceased. The creation of new elements was halted in its tracks, and the chemical composition of the Universe was frozen in at three-quarters hydrogen and one-quarter helium (with trace, and essentially ignorable, amounts of lithium). There it remained fixed for 200 million years. If the periodic table of the elements had existed back then, it would have consisted of only two tiles.

At first, the hydrogen and helium existed as wisps draped across the new-born Universe, dispersed so thinly they were barely there at all. Everything was cold and dark because there were absolutely no stars shining forth heat and light.

After those first 200 million years, pockets of hydrogen and helium began to clump together. The first nebulae were slowly accumulating. Having no stars set in their midst, the earliest nebulae did not glow incandescent as they do today, but within them, the seeds of starlets were being sown. Clumps of gas began to collapse inwards under their own gravity to form dense spheres shrouded in an orbiting disc of swirling wisps. In the core of the dense spheres, temperatures and pressures reached such soaring heights that fusion reactions were ignited, and the giant orbs of gas flickered into light.

Thus, the first stars were born, and the Universe was flooded with light. The long days of darkness were over; the days of starlight had begun.

In the gaseous discs orbiting the first stars, the first systems of planets coalesced, but these early solar systems were quite unlike the one which we inhabit. Because there were still only two elements in the entire Universe (hydrogen and helium), no dust could form from the protoplanetary discs as they cooled. The steady condensation of minerals, from gas to dust, could not happen. Chemistry was not yet a feature of the Universe, and so rocks could not exist. The first planets were made only of gas.

But things were slowly changing.

Cosmic crucibles

Our Sun is an average-sized star, but even so, it weighs an aston-ishing two million billion billion billion kilograms. With such masses, stars have intensely powerful gravitational fields, and the weight of overlying hydrogen and helium inflict astronomical temperatures and crushing pressures on their centres. Here in the stellar cores, hydrogen atoms fuse to create helium by a series of nuclear reactions. Over billions of years, element number one is gradually forged into element number two.

Enormous amounts of energy are released when hydrogen nuclei combine to form helium. The creation of new elements is the reason that stars shine. It is the reason that the Earth is not a cold dead place, but basks in a warm glow. The Sun is a life-giving fusion reactor.

As the hydrogen in a star's core is slowly consumed, the accu-mulation of helium begins to hamper the fusion reactions like a fire being choked with ash. Eventually, the hydrogen fuel runs out and fusion reactions cease. The star, for a brief moment, stalls, and, unable to hold itself up against its own crushing gravitational pull, contracts.

Fresh hydrogen is drawn closer to the helium-choked centre and, like dry tinder pressed to a hot ember, it ignites once more. The star is by now comprised of a white-hot helium core veiled by a shell of dancing hydrogen undergoing nuclear fusion. Further inwards the inert core slumps, and further the temperature rises. The outer layers of the star, lofted by the intense heat from deep within, are blown backwards into space, and the star expands to a hundred times its original size.

This is the fate of our Sun. In some five billion years or so, it will exhaust the hydrogen fuel in its core and expand outwards into the inner Solar System, engulfing Mercury, Venus, and the Earth as it swells.

As the shell of burning hydrogen around the core steadily fuses into helium, it too begins to choke on helium ash. The inert ash-ridden core grows slowly larger and, in doing so, contracts more tightly and becomes hotter. Such temperatures are eventually reached by the crushed helium that it ignites in a sudden flash. The core begins to glow a billion times brighter than it did before and becomes host to nuclear fusion reactions once more. Nuclei of element number two, helium, come together in threes and fours to form elements number six and eight – carbon and oxygen.

As the first stars in the Universe evolved through these stages, chemistry was gradually coaxed into being.

Eventually, the helium-burning core becomes choked with carbon and oxygen ash, and stalls once again. A second collapse ensues. The steadily slumping core of an average-sized star never reaches the crushing pressures and temperatures to fuse carbon and oxygen, and so their cores flicker out and become forever inert.

Ferocious temperatures in the slumping carbon–oxygen core cause the outer shells of the star to be propelled backwards into space. In a gentle puff, the star disperses itself outwards in all directions. Fusion reactions cease entirely throughout its interior, marking the end of its ten-billion-year life. Pulses of extreme mass loss carry glowing hydrogen and helium backwards into the realm of interstellar space. New nebulae are born, and eventually, that gas will collapse once more to form stars anew. In the unending cycle of cosmic renewal, some of it will form new solar systems.

Beyond oxygen

Stars more massive than the Sun are not extinguished when their cores become choked by carbon and oxygen. They have such a high mass – so much matter pressing down upon their cores – that they are hot enough to burn onwards. The successive stalling of smothered cores initiates fusion reactions to progressively higher and higher elements, which, one by one, populate the periodic table. Each time the fuel in the core is extinguished, core collapse increases the temperature and pressure a notch, stoking the fires and turning the ash into an inferno once more. Heavier and heavier elements are forged in the flames of the stellar fire each time the core stalls and reignites.

Helium nuclei fuse with oxygen to form element number ten – neon. The cores become inundated with neon ash, slump inwards further, and then the neon begins to burn. Helium nuclei fuse with the neon to form element number twelve – magnesium. Individual fusion reactions generate a minute quantity of energy, but because stars are giant cosmic structures, the sheer number of fusion reactions happening in each passing second is enormous, and so they add up to power an entire star.

From burning magnesium comes element number fourteen – silicon; from silicon comes element number sixteen – sulphur;

from sulphur comes element number eighteen – argon. Elements number twenty, twenty-two, and twenty-four – calcium, titanium, and chromium – follow.*

By this point, a rich blend of chemical elements has blossomed inside a star, but the chemical cocktail – from hydrogen to chromium – is tucked away deep below the stellar surface, unable to do anything. If the freshly synthesised elements are to go forth and form matter – such as molecules, minerals, meteorites, and minds – they have to leave their parent star.

The most massive stars – those with eight times or more the mass of our Sun – sustain temperatures of such ferocity, and pressures of such crushing extremity, that they burn onwards through the periodic table to even higher elements. Eventually, the stellar core hosts 'silicon burning' and synthesises elements number twenty-six and twenty-eight – iron and nickel – which spell the beginning of the end. Neither is burnable by nuclear fusion, and so the internal heat engine in the heart of the star begins to spin down. After burning brightly for millions of years, silicon burning is so fierce that it lasts only for a single day. The sphere of freshly synthesised iron and nickel in the stellar core grows quickly to the size of the Earth.

Then the star cataclysmically implodes.

With no energy production to counterbalance the momentous force of gravity pulling inwards, the frail core collapses. Like a nuclear explosion in reverse, the core shrinks from the size of Earth – some 13,000 kilometres across – to a mere 100 kilometres – about the size of Yorkshire – in one second, and with no core to support the star's outer layers, they plummet inwards.

---

* In reality, this chain of building heavier and heavier elements – each time increasing by atomic number two – is just one of a few chains progressing in parallel. In combination, these chains forge a rich suite of elements and various isotopes, including the ones with odd numbers of protons.

During the one second it takes the star to crumple, the collapsing layers are accelerated to one-quarter the speed of light: that is 75,000 kilometres per second: a suddenness beyond ordinary comprehension. As the centre of the star collapses, temperatures of 100 billion °C burst forth, and then something absurd happens: the falling layers rebound off the imploded core and bounce back as a terrible shockwave.

The star cataclysmically explodes. We call these explosions supernova, and for a short time, they outshine entire galaxies.

Unbound from its own immense gravitational pull, the massive star tears itself to pieces, releasing monstrous amounts of energy as it disintegrates. In the days and weeks preceding its demise, a supernova shines with the brightness of 100 billion suns. They are some of the most spectacular displays of light in the Universe.

The rich blend of chemical elements forged inside the star by nuclear fusion – from hydrogen to iron and nickel – are ejected backwards by the shockwaves, and long filaments and swelling plumes of incandescent gas flower into the cosmos. Draped like glowing tapestries, the ejecta from a supernova explosion decorate the ocean of interstellar space.

Left behind in the epicentre of the supernova are the star's remains. The gravitational field strength of the collapsed core is so strong that electrons are pressed into the nuclei of atoms, and the two combine to form neutrons. A spinning core, made entirely of neutrons, is all that is left, and is under such pressure that it shrinks to around ten kilometres across, but so dense that a single apple-sized lump would have the same mass as thirty cubic kilometres of rock. We call these city-sized stellar corpses 'neutron stars', and they are some of the most extreme and exotic entities in the known Universe.

Renewal

But the death of a star is no tragedy, and we need not mourn a light going out in the Universe.

The first generation of massive stars to grace the Universe sent forth their elemental medley into the cosmos, which, propelled by stellar winds and supernova shockwaves, enriched their surroundings with heavy elements. Interstellar nebulae, previously made only of hydrogen and helium, were enriched by the chemical fallout. Cosmochemistry was born.

Atoms forged in the first generation of stars became mixed into nebulae and collapsed with them to form new solar systems. As they massed into swirling clouds collapsing under their own gravity, these chemically enriched nebulae, with their lavish suite of new elements, held much potential. The whirling clouds flattened to form protoplanetary discs that slowly cooled. Housing elements from across the periodic table – unlike the first generation of stars – the second generation of protoplanetary discs had a rich blend of elements with which to forge rocks and chronicle the geological history of worlds.

The elements that were fashioned inside massive stars were steadily crafted into new forms of matter inside infant solar systems. Phoenix-like, minute grains of rock – such as CAIs and chondrules – emerged from the ashes of deceased stars, and the first rocks in the Universe bloomed into existence. The geological record of the Universe began. The motes of nebula dust coalesced to form planetesimals, comets, and eventually, planets, just as they did in our own Solar System 4.6 billion years ago. Nuclear fusion in the heart of massive stars is a prerequisite for geology, and without it, the Universe would consist only of hydrogen and helium.

The many elements spanning the periodic table show that our own Solar System is made, at least in part, from stellar fallout. While our Solar System is old, it is nowhere near as old as the Universe. Stars had more than nine billion years to enrich the Universe with heavy elements before our parental nebula began collapsing, which was plenty of time for it to be fertilised with elements higher than helium.

In the 1970s, cosmochemists at the California Institute of Technology ('Caltech') were poring over giant CAIs plucked from pieces of the recently fallen Allende. In their careful work they discovered something entirely unexpected: anomalous excesses in the heavy isotope of magnesium, ^{26}Mg. Magnesium-26 is the daughter product of our old friend aluminium-26 (^{26}Al), which was the short-lived radioisotope responsible for melting the earliest asteroids. The only way to explain these slight excesses of ^{26}Mg was if there was 'live' ^{26}Al present in the Solar System as it was forming.

The discovery of short-lived radioisotopes like ^{26}Al proves beyond reasonable doubt that our nebula was seeded with stellar fallout right before it collapsed. The time between the synthesis of ^{26}Al in a star and its incorporation into our nebula must have been short, no more than a few million years, otherwise it would all have fizzed away to magnesium-26 by the time it reached us.

Synthesised inside a dying star; blown across interstellar space; mixed into our collapsing nebula; incorporated into CAIs and the first asteroids: the story of ^{26}Al unfolded rapidly.

Giant stars were shedding their enriched outer layers and popping like firecrackers all around the nascent Solar System. Ferocious winds from the surface of these stars, carrying their swarm of synthesised elements and radioisotopes, carved cathedral-like caverns into the surrounding nebulous clouds. The clouds were crafted into regions of high turbulence and density, and perturbations in the billowing gas caused pockets to give way to gravity. It is possible that these winds, emanating from stars in their death throes, actually caused the collapse of the nebula in the first place.

Runaway collapse ensued. New stars and protoplanetary discs, by now seeded with a cosmochemical medley, evolved into a new existence. At least one of them – our Solar System – would go on to host collections of chemical molecules capable of contemplating their own story.

Elements up to and including iron* are forged inside the burning hearts of massive stars by nuclear fusion. But this immediately raises an important question: where do the even heavier elements come from?

A casual chemical analysis of any meteorite – or any rock, for that matter – will reveal elements from across the entire periodic table, many of them well beyond iron, which is only element number twenty-six. The highest naturally occurring chemical element found on the Earth is element number ninety-two, uranium. Each of the sixty-six elements in between – everything from element numbers thirty-three (arsenic), forty-seven (silver), and eighty-two (lead) – cannot be made by nuclear fusion in a stellar furnace. There is something inside stars running in parallel to fusion.

The nebula from which our Solar System collapsed was enriched with aeons worth of chemical commodity from many long-dead stars. Elements heavier than iron are synthesised inside stars by two means: the 'slow-process' and the

---

* In reality, nickel-56 is the heaviest nucleus forged inside a star, but being highly radioactive, it decays into insignificance within weeks of forming. Therefore, the heaviest element on Earth produced via nuclear fusion is iron (specifically, iron-56).

(equally imaginatively named) '*rapid*-process'. Both processes work in similarly creative ways but play out over vastly different timescales, and, whilst some elements – or specifically, some isotopes of some elements – can be made by both processes, others are made exclusively by one or the other. Coincidentally, each contributes equally to the budget of elements heavier than nickel in our Solar System.

The *slow*-process and the *rapid*-process both hinge upon the addition of neutrons to atomic nuclei, and each requires the presence of chemical seeds – pre-existing high elements – in a star. This style of element building could not operate in the first generation of pure hydrogen–helium stars; it only works in stars that collapsed from a nebula already seeded with at least a smattering of higher elements.

At first consideration, it seems counterintuitive that new elements can be synthesised by the addition of neutrons. After all, it is the number of protons inside an atom's nucleus that defines the element, not the number of neutrons. While they are bystanders in the complex realm of chemistry, neutrons affect only an atom's mass, giving rise to the clutch of isotopes that occur for most elements.

But it is precisely the power to generate isotopes of the same element that gives neutrons the ability to forge new elements entirely. As we discovered in the case of ^{26}Al and the cabinet of natural atomic clocks, not all isotopes are stable, and when they radioactively decay, they turn into new elements. No fusion is required.

Building slowly with neutrons

Rather predictably, the *slow*-process is the slower of the heavy-element smiths.

Once a star's core has become choked with inert carbon and oxygen, copious amounts of neutrons are produced in the helium- and hydrogen-burning shells surrounding it. The neutrons are simply by-products of nuclear fusion. If these shells are lightly seasoned with pre-existing heavy elements – such as iron, forged in and inherited from a previous generation of massive stars – the *slow*-process may commence.

Every once in a while, an atom happens upon one of these solitary neutrons, and if it hits just the right spot at just the right speed, the two merge. Thus, its mass number increases by one and a heavier isotope is formed. But the element, for the time being, stays the same.

Neutron absorption in these shells is an arduous process. An atom may go a few hundred or a few thousand years between each encounter with a lone neutron (hence the name *slow*-process), but each time it does, it becomes a slightly heavier isotope of itself.

If a nucleus consumes more neutrons than it can handle, it can no longer hold itself together. This is where the magic of the *slow*-process happens: unstable isotopes undergo radio-active decay. A slight excess of neutrons in an atomic nucleus causes one of them to spontaneously decompose and transform into a proton. A proton now takes the place of a neutron. Since it is the number of protons that dictates an atom's elemental nature, this changes its essence entirely, and a brand-new chemical element is formed.

Painstakingly, as the aeons creep by, neutron after neutron after neutron is slowly absorbed by chemical seeds. Each time a radioactive isotope is forged, it quickly decays, increasing the number of protons by one and thereby forming a new element. One by one, heavier and heavier isotopes are built; from them come higher and higher elements. From the iron limit of nuclear fusion right up to and including element number eighty-two (bismuth), the *slow*-process of neutron capture unfolds and heavy elements are synthesised in an orderly sequence.

Take iron-56 (^{56}Fe), the limit of nuclear fusion, as an example. If a star contains ^{56}Fe fallout from a previous generation of stellar giants, then it will act as a *slow*-process seed. When ^{56}Fe absorbs a neutron, its mass number increases by one and it becomes ^{57}Fe; another 10,000 years may pass until another neutron is absorbed to create ^{58}Fe. Upon absorbing a third neutron, radioactive ^{59}Fe is forged, which swiftly undergoes radioactive decay to become cobalt-59 (^{59}Co). From iron to cobalt; from element twenty-six to element twenty-seven. The star overleaps the limit imposed by fusion.

As a star wanders down the winding path of deep time, the *slow*-process snakes its way up the periodic table, hammering out new elements one by one. Every time a radioactive roadblock is reached by successive neutron captures, the atom leaps up the periodic table to the next element, and slowly the star entices rich chemistry into being.

Building rapidly with neutrons

Whilst the *slow*-process forges higher elements at a leisurely pace inside almost every star, its twin, the *rapid*-process, is hasty, and demands a particular set of finely tuned circumstances.

The extent to which the *slow*-process can forge heavy isotopes of a particular element is limited largely by one factor: the radioactive decay rate of the newly formed isotope. Once a radioactive isotope is forged by the *slow*-process, unless it decays exceptionally slowly, it will perish before it has a chance to absorb another neutron. No such barrier stands in the way of the *rapid*-process.

There are stellar environments that host such extreme densities of neutrons — some 1,000 billion billion of them packed into every cubic centimetre – that the time limits imposed by *slow*-process do not apply. Because so many neutrons saturate these environments, the atomic seeds absorb neutron after neutron after neutron after neutron in a single fraction of a second. They simply do not have time to decay before another neutron is packed into them. These nuclei rapidly swell to grotesque caricatures of their lighter selves, and in doing so forge a medley of ultra-heavy isotopes.

Such neutron-rich environments exist only in the most extreme of stellar environments. They are, unlike the tame and level-headed conditions that give rise to the *slow*-process, fleeting features in the Universe, and neutrons flowing in such dense swarms last mere minutes.

Suddenly, the intense supply of lone neutrons is halted, leaving in its wake a motley assortment of ultra-heavy isotopes. These freak isotopic aberrations are incredibly unstable; they immediately begin tumbling down a waterfall of radioactive decay after decay, each time leaping up through the periodic table and creating a higher and higher element. Radioactive decay continues until a stable isotope is happened upon, and at that point, the cascade of decays abruptly ends.

Every element on the periodic table heavier than bismuth, up to and including uranium, is forged by the *rapid*-process in these most extreme stellar furnaces. This includes some of our most highly prized substances, like elements number forty-seven, seventy-eight, and seventy-nine: silver, platinum, and gold.

Cosmochemistry from afar

The presence of neutron-rich isotopes in our Solar System that were synthesised by the *rapid*-process – both in Earthly rocks and in meteorites – proves that our nebula was seeded with fallout from extreme neutron-rich stellar environments. But for a long time after the 1950s, when the *rapid*-process was first hypothetically formulated on paper,[1] exactly where it happens in Nature was confined squarely to the realm of speculation. It was clear that an incredibly neutron-dense stellar environment was required to synthesise a huge swathe of the periodic table; but where on earth (or rather, where in space) one could find some thousand billion billion neutrons packed into every cubic centimetre was a mystery. Such densities are simply not attained by normal stars.

One prime candidate for the source of *rapid*-process isotopes were stellar explosions: supernovae. After all, they outshine entire galaxies, and often create an ultra-dense neutron star at the explosion's epicentre during core collapse. Perhaps the newly fashioned neutron star is the potent source? It seems intuitive.

Computer simulations of supernova explosions and mathematical models describing the physics of neutron star formation repeatedly fail convincingly to reproduce the conditions needed to forge *rapid*-process isotopes. This is a problem. It seems that most of the neutrons in the neutron star are trapped in the remnant core and do not engage in the business of forging new isotopes.

For decades, physical observations shed no light on the situation, either. The cascade of radioactive decays before a neutron-rich non-radioactive isotope is reached releases enormous amounts of energy, which would cause the material ejected from a supernova to glow in a characteristic and predictable way. No such afterglow had been observed in the sky following a stellar explosion.

Supernovae, despite their appeal, are probably not the environment in which *rapid*-process isotopes are forged. Thus, the realm of *rapid*-process nucleosynthesis and the origin of half of the elements heavier than iron in our Solar System remained shrouded in mystery, but, tantalisingly, the veil was lifted by one of the most thrilling stories of modern science in recent times.

On 17 August 2017, after they had embarked on a 130-million-year journey across intergalactic space, wrinkles in the fabric of the Universe arrived in our Solar System. They rippled through the Earth before continuing their light-speed voyage through the cosmos. These small perturbations in spacetime, detected simultaneously at two astrophysics observatories – the Laser Interferometer Gravitational-Wave Observatory (LIGO) in Washington (USA) and the Virgo Interferometer in Cascina (Italy) – lasted for around 100 seconds; we call such undulations 'gravitational waves'. The observation of this 100-second period of spacetime choppiness changed everything.

Gravitational waves are creases in the fabric of spacetime. They are sent forth across the Universe when two inconceivably massive objects collide and merge. They were a key prediction of Albert Einstein's General Theory of Relativity (1915), although Einstein believed that they would be too feeble ever to detect. But in 2017, we did detect them.

The gravitational-wave detectors at LIGO and Virgo can be thought of as gravitational telescopes. Comprised of a four-kilometre-long L-shaped tube housing a complex series of lasers, each is capable of detecting ripples in spacetime of one-ten-thousandth the diameter of a proton (0.0000000000000000001 metres). This is the mind-bending level of precision it takes to detect fleeting waves in the fabric of the Universe.

Something like 130 million years ago in a distant galaxy (while dinosaurs were walking on the surface of the Earth), two ultra-dense stellar corpses – neutron stars – were locked in a death spiral. Orbiting one another in an intricate pirouette, the neutron stars gradually leaked away their sustaining gravitational energy into space, and so they slowly slipped closer together.

Inch by inch they spiralled closer, until, quite suddenly, they merged. The last throes of their gravitational energy were expelled from the dancing duo in the form of gravitational waves; the same gravitational waves we humans would detect on 17 August 2017 came from colliding neutron stars.

Immediately after the waves were detected, a worldwide alert was issued to the astronomy community to search for where they came from in the sky. And the world responded. Within eleven hours, a team of astronomers using the Swope Telescope in Chile had found the source.

The two merging neutron stars – each more massive than the Sun yet only the size of a small city – had released a truly eye-watering amount of explosive energy. The fallout from the detonation ballooned at one-quarter the speed of light and sparked a spectacular light show that shone out across the Universe.

Teams of skywatchers from across the globe, using both ground- and space-based telescopes, launched an extensive observation campaign that would last for months. Light spanning the electromagnetic spectrum – from radio waves, infrared, optical, and ultraviolet, to X-rays and gamma-rays – was collected from the explosion; more than 3,600 scientists from across the planet took part in the unprecedented effort to study this single celestial firework.[2]

The two neutron stars had violently torn themselves to shreds in their final seconds and blasted their surroundings with copious quantities of neutrons. The fallout from the explosion was humming with them. In the days following the blast, the rapidly expanding shower of neutrons and superheated gas cooled, but remained hotter than one would expect. Something was keeping it warm.

Some of the isotopes in the fallout were clearly radioactive. As they decayed, they released energy into their billowing surroundings, keeping the fallout from cooling too swiftly. The atomic species inside the mushrooming ejecta also absorbed some of the light in the afterglow, but in a systematic and orderly way: different chemical elements absorbed specific wavelengths of light. Astronomers used these spectroscopic fingerprints to unpick the chemical make-up of the burst and discovered it to be brimming with heavy elements. Even now, years after the explosion, astronomers are still combing through the spectroscopic data and identifying new elements in the fallout (for example, strontium – element number thirty-eight, which features in our bones and teeth – was identified in late 2019).

One hundred billion billion billion kilograms of heavy elements were ejected from this single explosion. That is some 15,000 Earths worth of matter. Of the 15,000 Earth masses, around ten of them were gold; the explosion quite literally spun gold into existence and sent it reeling forth into the cosmos.

Ample neutrons; radioactive isotope fallout; heavy elements. These lines pointed towards fallout that was chock-full of atoms forged by the swift absorption of neutron ejecta, which were briskly decaying to form neutron-rich isotopes of stable elements. The astronomers had done it. They had found the birthplace of half of the chemical elements in the Solar System heavier than iron, and the crucible that cooked up all of the elements heavier than bismuth. They had found the site of the *rapid*-process. Theoretical astrophysics and the composition of meteorites had raised the question of where these elements and isotopes came from, and the gravitational waves illuminated the path towards an answer.

Aside from the primordial wisps of hydrogen and helium that were forged in the minutes preceding the Big Bang, every element on the periodic table was, by and large, synthesised inside stars.* Elements lighter than iron were brought into existence by fusion reactions; those heavier were assembled largely by the absorption of neutrons during the *slow*-process and the *rapid*-process.

Elements blown across the cosmos in the form of gas and dust by dead and dying stars were stirred into our nebula, and the atoms formed in different stellar environments were mixed to create the cloud from which our Solar System formed.

Aside from a slight variation from place to place, our Solar System is remarkably uniform. True, there are more ice-rich bodies further away from the Sun in the cold regions of the outer Solar System; and yes, there are variations among the asteroids, such as their precise blend of oxygen isotopes. But such variations were largely inherited from the physical and chemical evolution of the protoplanetary disc after it collapsed.

Any isotopic variations inside the nebula, which reflected the numerous stellar sources that contributed the atoms making up our Solar System, were largely erased while it was churning prior to its collapse. The intense heat generated during the formation of the disc vaporised and mixed almost everything, including smatterings of dust that collapsed along with the gas. We lost the scintillating details of our rich stellar ancestry. The nebula, as it existed just before the CAIs, chondrules, and asteroids began forming, was almost flawlessly uniform.

Almost.

* Except for elements numbers four and five – beryllium and boron – which are not made inside stars. They are only produced when atoms are cleaved into pieces by high-energy cosmic ray bombardment.

Strange noble gases

By the middle of the twentieth century, hints of strange and extreme isotopic signatures hidden deep inside meteorites were beginning to emerge, and some of the first elements in which bizarre isotopic signatures were discovered were the noble gases. These elements – helium, neon, argon, krypton, and xenon – occupy the right-hand column of the periodic table and are famed for their inertness. They have little tendency to participate in the complex business of chemical reactions and crystal growth, and so when they exist inside rocks, they often do not sit as part of a rigid mineral structure *per se*. Rather, they perch uncomfortably inside crystal arrangements, trapped between other atoms.

Noble gases also have plenty of isotopes: helium ('He') has two; neon ('Ne') and argon ('Ar') each have three; krypton ('Kr') has six; and xenon ('Xe') tops them all with an impressive nine. Each isotope is forged by a different combination of stellar synthesis. For example, the middle five isotopes of xenon – ^{128}Xe, ^{129}Xe, ^{130}Xe, ^{131}Xe, and ^{132}Xe – are manufactured in part by the drip feeding of neutrons during the *slow*-process; the heaviest two – ^{134}Xe and ^{136}Xe – are manufactured exclusively by colliding neutron stars in the *rapid*-process.

Some of the first puzzling measurements came from xenon. By gradually heating small chunks of meteorites – chondrites, specifically – in a vacuum furnace devoid of air, cosmochemists in the mid-1960s progressively broke down and released tiny puffs of trapped gases from the stone, xenon amongst them. Carefully, they funnelled the liberated xenon into a mass spectrometer and carefully measured its precise isotopic make-up.

The isotopic blend of the trapped xenon released at temperatures approaching 900 °C was like nothing that had been measured in any rock, ever, from anywhere in the Solar System. By their nature, chondrites are an assembly of innumerable individual motes of nebula dust that formed at different times and places in the protoplanetary disc. There was something inside these meteorites – some as yet unknown type of cosmic sediment – that housed the strange xenon and released it upon reaching red-hot temperatures.

Similarly freakish isotope anomalies were soon found for trapped neon. Krypton, nitrogen, and carbon followed. Their exotic isotopic blends defied explanation. Isotopic variance across the Solar System usually amounted to less than one per cent of one per cent, but these ratios differed from the norm by factors of many hundreds. There was no known chemical or physical process in our Solar System that could fabricate the bizarre isotope compositions scattered inside the chondrites, especially if the Solar System was well mixed and uniform when it formed.

Isotopic needles in cosmic haystacks

Guided only by the vague beacons of strange noble gases, cosmochemists began the search for the source of the isotopic eccentricity. Microscopes were of little use because the nature of the anomalous cosmic sediment was entirely unknown.

An entirely new approach came in the mid-1970s when cosmochemists at the University of Chicago were foraging for anomalous grains in chunks of Allende. Having fallen only a few years previously and being chock-full of pristine cosmic sediments such as CAIs and chondrules, Allende was the perfect meteorite for the task.

Large chunks of Allende were placed into a chemical-resistant vial and dissolved in some of the most aggressive acids known: vigorous hydrochloric acid; fuming nitric acid; deadly hydrofluoric acid; bubbling *aqua regia* (Latin for 'royal water'), a furious mixture of concentrated hydrochloric and nitric acid, famed for its ability to dissolve even gold. The boiling acids tore Allende to pieces on an atomic level and, like sugar being stirred into hot tea, the meteorite disintegrated. Minerals that had survived 4.6 billion years of Solar System history – CAIs and chondrules among them – were entirely destroyed.

But remarkably, even against the roiling tide of caustic reagents, some minerals survived.

A thimbleful of pale microscopic crystals sat in the bottom of the vial, entirely unscathed by the harsh chemistry. These grains were incredibly tough. Carefully, the cosmochemists decanted the acidic solution from the vial, leaving behind only the microscopic grains, which were swept up and placed inside a mass spectrometer equipped with a furnace. Gently, the cosmochemists relieved these strange grains of their trapped noble gases by ratcheting up the temperature and cooking them.

They were a perfect match. The nanoscopic Allende grains, impervious to the corrosive power of the acids, were the source of the strange isotope signatures. Large chunks of Allende were sacrificed during the endeavour, but by burning down the haystack to find the needle, cosmochemists had at last isolated the bizarre cosmic sediments that hosted the isotopic anomalies.

In the 1980s, more anomalous grains were isolated from a suite of chondrites – Allende included – by acid digestion, and this time they were placed under the lens of a powerful microscope rather than inside a mass spectrometer. On the boundary between the geologic and the atomic, some of the grains were only five nanometres across (1,000 times smaller than a single red blood cell). Despite their tiny size, the isotopically strange grains were identified; their mineralogical character shocked the cosmochemistry community. They were diamonds. Real, crystalline, cosmic diamonds.

More anomalous grains were isolated and characterised in the following few years. Additional nanodiamonds were discovered, alongside nanoscopic crystals of graphite (the non-diamond form of carbon) and silicon carbide (an exceptionally rare mineral that is found almost exclusively in meteorites); all three of them housed extreme blends of isotopes in every element it was possible to measure. By dissolving chunk of meteorite after chunk of meteorite, the inventory of these tiny isotopic oddballs grew and grew.

But why were these nanoscopic pieces of rock so different from the 'normal' matter from which our Solar System is forged? What lay behind their vastly abnormal isotopic character? No known process that operates in our Solar System – physical or chemical – can fabricate isotopic anomalies of this magnitude.

And therein lies the answer. No known process that operates *in our Solar System* can fabricate isotopic anomalies of this magnitude. These grains could not have crystallised from the well-mixed protoplanetary disc in orbit around our young Sun alongside the CAIs, chondrules, and other 'normal' motes of cosmic sediment.

The answer is inescapable. They crystallised elsewhere. They crystallised around other stars. These grains are pieces of other solar systems that survived the formation of our own planetary system, and persisted within meteorites for 4.6 billion years. They are pieces of bona fide stardust.

Stardust

In the winds emanating from dying stars, the fallout from supernova explosions, and the ejecta from colliding neutron stars, microscopic minerals condensed from gas to interstellar dust. The minute grains of stardust inherited exotic blends of isotopes from their parental star – signatures of fusion, the *slow*-process, and the *rapid*-process – before crossing the expansive ocean of interstellar space like wind-blown mariners.

In due course, they happened upon the nebula from which our Solar System would eventually form. Arriving in droves, the stardust mixed with the slowly turning filaments of incandescent gas, and there drifted through the clouds for aeons. Eventually, the nebula contracted – perhaps triggered by shockwaves emanating from a nearby supernova, which would have delivered more sprinkles of stardust – and collapsed to form the protoplanetary disc. The grains collapsed alongside it.

Most of the crystalline stardust was annihilated by the energy liberated by the collapsing disc. Upon vaporising, the grains became part of the gaseous cloud, blending into the thin wisps of nebula. Their isotopic strangeness was smeared out and lost for ever as it mixed with the rest of the gas. From gas to dust, from dust to gas, the ghosts of stars long dead evaporated into the breeze.

But some grains, against all odds, persisted. Ejection from their parental star and a journey across interstellar space; incorporation into our nebula and its collapse into a spinning protoplanetary disc; coalescing alongside the other pieces of cosmic sediment – the CAIs, the chondrules, and miscellaneous fine-grained matrix – to form the planetesimals; asteroid collisions and bombardment on a cataclysmic scale; 4.6 billion years in orbit around the Sun in the Asteroid Belt; ejection from their parent asteroid by an impact; an interplanetary journey to the Earth tucked inside a meteorite before a fiery plunge through Earth's atmosphere; only to face an onslaught of corrosive acids in cosmochemistry laboratories; and finally, to be miraculously discovered by the curious inhabitants of Earth. They survived it all. That they exist at all is nothing short of phenomenal; that we found them is a testament to the elegance of the scientific method and the storytelling power of meteorites.

By necessity, these minute pieces of stardust must have existed before our nebula began collapsing 4.6 billion years ago. They must pre-date our Solar System. This makes these tiny grains of stardust the oldest objects we can hold in our hands (though admittedly, they are invisible to eyes unaided by a powerful microscope). They are older than anything in our stellar neighbourhood, far pre-dating the first minerals to condense from our protoplanetary disc – the CAIs – and the Sun itself. In 2020, a team of cosmochemists led by the Field Museum in Chicago (USA) dated a suite of stardust grains using a natural isotopic stopwatch;[3] some were more than seven billion years old, pre-dating the Solar System by over three billion years. Tiny pieces of rock that are *seven billion years old*! The mind boggles.

Accordingly, we call these most remarkable motes of cosmic sediment 'pre-solar' grains.

We will, of course, never know exactly which individual stars they came from – most of those stars are by now long dead, anyway – but the pre-solar grains are some of the only physical proof that they existed at all, and they rain from the skies trapped inside meteorites.

For almost the entirety of human history, we could watch stars with the naked eye only; for the past four centuries, we have gazed at stars through telescopes; and now, by virtue of the pre-solar grains, we can see stars down a microscope. An alliance between two scientific instruments that at first seem entirely mismatched – the telescope and the microscope – has cast the stars in an entirely new light. We have tiny pieces of them here on Earth.

Our ancestors sensed that the stars were important, but had no way of knowing just how important, or to what extent they played a tangible part in the history of their own fleeting lives. We are fortunate to know what they could not.

Nucleosynthesis – the origin of the chemical elements and their medley of isotopes – is one of the most profound of all scientific discoveries. On a practical level, it answers a fundamental question: where do the chemical elements come from? On a human level, it answers a question we have asked for millennia: why do the stars shine? On a spiritual level, it is an integral part of that most timeless of questions: where do we come from?

Take a look around. The carbon in your hands; the cool nitrogen filling your lungs with each breath; the oxygen locked away in rocks beneath your feet; the fluorine, magnesium, and calcium dissolved in the water you drink; the iron flowing through your bloodstream; the heavy elements that we use to decorate our bodies, power our technologies, and craft our building materials. All of it was forged by nuclear reactions inside the white-hot interior of stars, the explosive demise of stars in their death throes, and the powerful collision of neutron stars. Much of this we have learned directly from meteorites.

Cosmochemistry is a wonderful and often bewildering field of science. It expands the horizon of human knowledge across multiple dimensions over scales of incredible magnitude simultaneously: from split second supernovae to aeons; from single stones to entire worlds; from microscopic crystals of dust to colossal stars. It never ceases to amaze me that deep insights into the entire Solar System can be glimpsed from a single pinhead-sized stone. Meteorites are capable of stretching the bounds of comprehension from the minuscule to the immense, and nowhere was this exemplified more beautifully than in the discovery of stardust tucked away inside the chondrites.

If the collapse of the nebula and the formation of the Solar System is chapter one of the human story, then the synthesis of the chemical elements by the stars is its prologue. If we follow the winding thread of deep time, backwards through geological time past the cosmochemical, we eventually arrive at the astronomical; and the pre-solar grains guide our way. They extend the rock record outwards across the stars.

Those shining lights in the sky are as much a part of our story as the collapse of our nebula, the condensation of nebula dust, and the formation of the Earth itself; it is all threaded together along an unbroken chain of events. And some of the atoms from which our Solar System is made, which ultimately saw their origin in stellar crucibles, evolved to the point where they became conscious beings that began asking questions about their origins. Us.

Some meteorites encompass a tale of origin a little closer to home than the formation of isotopically anomalous grains around remote stars. Chondrites, the unmolten aggregates of cosmic sediment, preserve the dusty building blocks of worlds, and a subset of them – the carbonaceous chondrites – contains something equally as remarkable and altogether more rousing: the chemical building blocks of life itself.

8

STAR-TAR

It was 1969.

Allende, the two-tonne meteorite brimming with pristine cosmic sediments, had fallen from the skies over Mexico in February. From this single meteorite came a multitude of discoveries: the Sun-like oxygen isotope composition of the CAIs; the presence of 'live' ^{26}Al in the nascent Solar System; the isolation and characterisation of pre-solar stardust made from diamond; and the precise age of the Solar System — 4,567 million years — by dating CAIs with the uranium–lead isotope clock. Allende remains one of the most intensely studied rocks in the history of science, but its fall was just the beginning of what 1969 had to offer.

In June the *Apollo 11* astronauts journeyed to the Moon aboard a Saturn V rocket. Leaving the first human footprints on the Lunar surface behind them, they brought over twenty kilograms of Moon rock back to the Earth.

And on 28 September in a quiet farming village in Victoria (Australia), something happened that would forever change the way we see our place in the Solar System. It centred around the arrival of another celestial stone.

A congregation

Clear skies graced the village of Murchison and were pervaded by a languor only Sunday could bring. The air was still and hot. Locals were getting ready for church: shoes were being shined, hair brushed, clothes pressed, when, suddenly, the tranquillity was broken.

An explosion was followed by a roar that seemed to emanate from the Earth and sky simultaneously. It was as though a jet aeroplane had taken off in the village. Cattle, shaken by the terrible blaring, rushed frantically about their paddocks, and the locals were jerked out of their Sunday morning slowness. The oppressive noise seemed to bear down from all sides.

Echoes of a second detonation rang across the landscape and the roaring sound reached new heights. Then came a light. Blinding orange fire surrounded by a white-hot halo tore across the skies above Murchison, outshining the mid-morning sunshine. The trail of fire in its wake quickly quenched to a streak of blue-grey smoke and was spotted over 400 kilometres away. A third explosion. Then the rumbles dissolved back into the air and silence returned.

Speculation ran rife among the church congregation, with explanations ranging from the tragic – *colliding aeroplanes* – to the fantastical – *fallen spacecraft debris* – to the supernatural – *alien spaceships battling across the skies*. Whatever it was, nobody had ever seen anything like it.

Beneath the dwindling trail of smoke left behind by the plummeting fire, fallen stones stippled the ground. Miraculously, the only obvious damage was to a barn where a fist-sized piece had crashed through the roof and landed atop the hay. The stones were pitch black and covered in a smooth, varnish-like crust. A meteorite had just fallen over Murchison.

Within the hour, villagers were feverishly collecting pieces, and it quickly became clear that their apparent celestial origin was not the strangest thing about them. There was something odd about these particular rocks. They *smelled*.

A strong chemical aroma, as sharp as paint stripper, emanated from the stones. Some villagers were initially wary, afraid that the pungent fumes were poisonous. Curiosity eventually got the better of many, and close to 100 kilograms of rock were collected from across the farmland and squirrelled away in people's houses. Crucially, the stones did not have time to be afflicted by atmospheric phenomena – like rain – before they were picked up. They remain to this day some of the most pristine meteorites ever collected.

Within five days of the fall, news of the spectacular fireball and pieces of the strange stone had made their way to Melbourne University's Geology Department. A few fragments came into the possession of Professor John Lovering, the Head of School. Lovering noticed that the meteorite's fusion crust had fallen away in places to reveal its interior: coal-black rock peered outwards from behind the veil. Set amongst the dark, sooty matrix were tiny flecks, some snow white and fluffy, others circular and grey, reminiscent of granular sedimentary rocks so common here on Earth. This was cosmic sedimentary rock: the fluffy white stipples were CAIs and the round beads were chondrules. It belonged, like Allende, to a rare class of meteorite: the carbonaceous chondrites.

But in more ways than one, the Murchison meteorite was different from Allende. Like the Murchison villagers, Lovering too noticed the strange stench emanating from the stone: 'When I first saw the meteorite, I found it was in a plastic bag. And when it was opened up, suddenly this great organic chemistry smell hit me, just like methylated spirits: very, very strong!' As one might expect, meteorites that smell are exceptionally uncommon.

'This [meteorite] is almost as exciting as moon dust!' Lovering declared to the press. He was holding what would go on to be one of the most intensely studied pieces of rock in the history of geology and cosmochemistry, rivalling even Allende. In many ways, the Murchison meteorite would turn out to be more exciting than pieces of the *Apollo* Moon rocks.

Before we delve into what was uncovered beneath Murchison's fusion crust, let us briefly explore an element that played a pivotal role in bringing the study of meteorites irreversibly into the scientific limelight: carbon.

The element of life

Carbon is the fourth most abundant element in the Universe, and by mass is the second most abundant in the human body after oxygen. It has a combination of chemical properties that give it a proclivity to bond easily with other chemical elements – including other carbon atoms – enabling it to form an immense array of molecular compounds. Carbon is so well-suited to bonding with other elements that there is an entire branch of chemistry centred around it: 'organic chemistry'.*

Tens of millions of organic molecules have been identified and characterised by chemists over the past century. Organics exist in many shapes and sizes. Some comprise long chains of carbon atoms on which smaller groups of atoms are threaded like a string of pearls, often encompassing (but not limited to) hydrogen, oxygen, and nitrogen, and to a lesser extent phosphorus and sulphur. Others consist of a central ring of carbon

---

* Not all carbon-bearing molecules are considered organic, e.g. carbon dioxide (CO_2). Precisely what defines a molecule as organic lacks consensus amongst chemists.

atoms adorned with protruding molecular groups like twigs jutting from a wreath; and others are an intricate combination of the two, featuring chains and rings banded together into molecular metropolises constructed from hundreds of atoms. All organic molecules have one thing in common: a skeleton of carbon.

Carbon is, so far as we know, absolutely central to life, earning it the nickname 'the element of life'. Every single living entity that we know of is carbon based. Alongside hydrogen, oxygen, nitrogen, phosphorus, and sulphur, carbon makes up the bulk of living organisms, from oak trees, to yeast, to trees, to blue whales. More than ninety-six per cent of the mass in your body is carbon, oxygen, hydrogen, and nitrogen alone.

Despite their ubiquity in our everyday experience of the world (as living organisms ourselves), organic molecules are rather uncommon on the Earth on a planetary scale. They are scarce in the rocks beneath our feet. Even when they are found scattered inside the geological fabric of a rock – like the hydrocarbon fossil fuels that power our civilisation: coal, oil, and gas – nearly all organics ultimately arise from a once-living source. Oil, for example, is itself derived from the decomposition of deceased organisms that were buried beneath layers of sediment in the depths of the geological past.

Despite the exceptional rarity of organic molecules in the subterranean rock record, we do occasionally find them in one of the most extraordinary and unexpected places: meteorites.

Pride rock

By the mid-nineteenth century, the notion that stones really do fall from the sky was widely accepted throughout the scientific community. This was fortunate, because on 14 May 1864, a particularly unusual rock made landfall in France, and was diligently collected before it had a chance to be consumed by the effects of terrestrial weathering.

At thirteen minutes past eight in the evening, the fireball flooded the French city of Montauban with a brilliant luminance. Onlookers recalled it as being: 'as large as the full Moon that crossed the sky as a shooting star would'.

Light danced for 500 kilometres in all directions, and deep rumbles from the sonic boom rolled across the countryside of southern France and into the northern regions of Spain. The spectacular event culminated in the fall of stones just south of Montauban, which marked the 219th witnessed meteorite fall recorded in history (and the fortieth on French soil). News of the celestial light show quickly reached the ears of the French scientific establishment. Close to fifteen kilograms of rock was quickly recovered from across the region, and the meteorite was named after the village around which it fell: Orgueil (also the French word for 'pride').

Most of the meteorites known at the time were, for lack of a better word, rocky. If somebody with a good aim were to throw one at you, the meteorite would survive the ordeal unscathed. (You, however, would not be so lucky.) The iron meteorites were made from metal, after all, and so could withstand even the most careless handling, but even the stony meteorites (the stony achondrites and the chondrites) were still solid.

Orgueil was different. In many ways it resembled a friable lump of coal rather than a proper rock: it was pitch black both outside and in, dotted only here and there with minute paler flecks, and was so crumbly that a firm grip easily pulverised it. Upon exposure to water, the meteorite quickly disintegrated into what was described at the time as 'mud, black as shoeshine'. It is a miracle Orgueil survived its fall to the Earth's surface at all.

The first scientist to scrutinise the meteorite was acclaimed French chemist François Stanislas Cloëz. Within three weeks of the fall, Cloëz had discovered something remarkable inside Orgueil: a cocktail of complex organic molecules. He found similarities between the organics in Orgueil and bituminous rocks on Earth that contained the fossilised remnants of ancient lifeforms, and he determined that something like three per cent of the rock was made from carbon in one form or another.

Cloëz also reported the presence of water. Water! It was not possible to wring water from the stone as though it was a wet flannel; it was only through heating the stone that the water was liberated, indicating that it was bound within the structure of its constituent minerals. Cloëz reported that an astonishing ten per cent or so of the stone's mass came from water alone.

Orgueil was immediately classified as belonging to a recently founded and unusual group of meteorites, of which only five other specimens existed: the carbonaceous chondrites. It bore a striking resemblance to two previous carbonaceous falls in particular: Alais, which fell in France in 1806, and Cold Bokkeveld, which fell in the Western Cape of South Africa in 1838. All three of these meteorites were pitch black and held significant quantities of water and carbon, and each was soft enough to be easily splintered with a bare hand.

All three stones smelled upon being heated, too, emitting a pungent aroma reminiscent of oil. Each was practically brimming with organic molecules.

The collection of meteorites that hosted organic molecules and water was growing. It is difficult to imagine how excited and perplexed the cosmochemists of the day must have been to discover these substances in meteorites. At the time, organic molecules were thought to be exclusively the products of living organisms, and so their presence in stones that hail from heavenly bodies invoked the seemingly impossible. In a 1860 publication in which he described the organic molecules woven through the rocky fabric of Cold Bokkeveld, esteemed chemist Friedrich Wöhler opined: 'Based on our current knowledge, this organic substance can only have formed from organised bodies.'

Indeed, by the term 'organised bodies', Wöhler articulated the word on the tip of everybody's tongue: life.

The seeds of doubt

Throughout the remainder of the nineteenth century and into the middle of the twentieth, organic-rich water-bearing carbon-aceous chondrites continued to arrive on Earth.

Carbonaceous meteorites, while illuminating the story of our Solar System through a geological lens, were hinting at tales of a far closer and more intimate nature – the story of biology. The presence of the molecules of life in stones that originated from asteroids led many cosmochemists to speculate that these stones played a role in the origin of life on Earth, and may even hold hints that life exists elsewhere in the Solar System.

By the mid-twentieth century, even stones which sat on the ground for only a few days or hours before they were collected – like Orgueil – had still been inside the Earth's biosphere for close on a century. Many cosmochemists could not shake off the suspicion that many of the organic molecules, or perhaps even most, were the products of contamination from the lifeforms that inhabit every crevice of the Earth's surface, rather than having an otherworldly origin.

In a search for answers, they homed in on the most carbon-rich meteorites: the 'CI chondrites', named after the Tanzanian meteorite Ivuna that fell in 1938. To this day, only nine CIs are known (five falls and four Antarctic finds). They total just over twenty kilograms: Orgueil, at just under fifteen kilograms, easily makes up more than half of their total mass.

But Orgueil fell more than a century before sophisticated cosmochemistry curation facilities were developed, such as the one built at the Johnson Space Center to house the Antarctica meteorites and *Apollo* Moon rocks. Meteorites that fall in modern times, like Allende, can be rushed away into the sterility of a laboratory within days of falling. They are granted a steady flow of cool air, strict humidity controls, and quarantined from dirt, dust, and gloveless hands. Historic falls are less fortunate. They are often stored inside wooden museum cabinets or glass display cases, with little or no control over the conditions in which they sit.

After falling, most pieces of Orgueil wound up in the great museums of Europe, and there they were slowly infiltrated by the Earth's atmosphere and, inevitably, a small army of microbial life. One fragment, named 'No. 9419', arrived at the Musée d'Histoire Naturelle in Montauban, France, sometime between two and four weeks after falling. For ninety-eight years, No. 9419 sat patiently sealed inside a glass display jar, until it was cracked open by a team of cosmochemists wishing to study the organic molecules it contained. During its residency in the glass jar, the stone had crumbled somewhat from moisture in the air, illustrating the fragility of these carbon-rich meteorites.

The dark stone, partially covered by a varnish of black fusion crust, was flown to a laboratory in Chicago and carefully broken into pieces. The cosmochemists examining the stones immediately noticed something bizarre: there were tan-coloured pellets embedded in the jet-black matrix, unlike anything ever seen before in any meteorite.

They were not CAIs. They were not chondrules, either. They were seeds. And they were *inside* the meteorite.

The team thought they may have stumbled upon the ultimate discovery: organised bodies or, less euphemistically, life. Orgueil was a meteorite buzzing with organic molecules and practically soaking in extraterrestrial water. If the evidence of non-Earthly lifeforms were to be found anywhere, surely it would be in a meteorite like this one. Had they just discovered alien seeds?

The discovery of extraterrestrial life would change everything, and would require the most rigorous set of evidence. The team of cosmochemists in Chicago diligently deployed their full sense of scepticism and scientific prowess on the problem, and it did not take long for cracks to appear.

Along with the seeds, fragments of a familiar rock were found embedded into Orgueil: coal – the exact stuff that is used as fuel. They had previously gone unnoticed, blending in perfectly with Orgueil's blackness. This piece of Orgueil was riddled with these fragments.

A few months later, Albert Cavaillé, then Head of the Montauban Museum, positively identified the seeds as those of *Juncus conglomeratus*, a rush that inhabits boggy fields and pastures. *Juncus* is common in the south of France. The meteoritic seeds were not extraterrestrial: they were French.

The possibility of foul play developed into outright suspicion at a third discovery: glue. Derived from horse carcasses, the glue laced the inside of the meteorite, holding it together and fixing the seeds and coal fragments in place. Then it transpired that the 'fusion crust' was not fusion crust at all! Rather than forming when the exterior of Orgueil melted as it plummeted through the atmosphere at hypersonic velocity, this 'fusion crust' formed when the glue was painted onto the surface of Orgueil with a brush and left to dry into a shiny lacquer.

A great hoax was uncovered, and an account of it was published in the journal *Science* under the eye-catching title 'Contaminated Meteorite'.[1] A trickster – sometime in the weeks between Orgueil's fall and its arrival at the Musée d'Histoire Naturelle in Montauban in 1864 – had attempted to swindle scientists by weaving seeds of deception into the stone. The identity of the culprit, and his or her motive, have been lost in the depths of history, but it is at least possible that they were inspired by Cloëz' discovery of organic molecules. Whether they were trying to bolster speculation of life inside the CI chondrites or discredit the idea is not known.

Heavenly feet vs earthly thumbprints

The field of extraterrestrial organic chemistry saw a sharp resurgence in the early 1960s with the newfound reality of human space exploration. Scientists from the fields of geology, chemistry, and biology united, and put their minds to understanding one of the most timeless of questions: where does life come from? It is a question that every culture, religion, and mythology has attempted to answer.

How could it be that inanimate atoms – carbon, oxygen, hydrogen, and smatterings of other elements – which themselves are definitely not 'alive', combine to form beings such as us, who definitely are 'alive'? From such simple beginnings – from the swirling cloud of gas and dust in the protoplanetary disc – how did chemicals attain consciousness and become animated pieces of a planet? Put simply: how, when, and where did life arise? The processes that initially transformed non-living chemical elements into living organisms is formally known as 'abiogenesis'. Whilst there is no consensus among the scientific community as to exactly how, when, or where abiogenesis happened, there is no question that it did – otherwise, how could we be?

It is possible that part of this story played out in the protoplanetary disc and was recorded in the rock record of the infant Solar System. Carbonaceous chondrites may hold some answers.

Notwithstanding the Orgueil hoax, the origin of the complex organic molecules in some carbonaceous chondrites remained a credible and active area of scientific enquiry. There were three possibilities: that they were from Earth's biosphere and were little more than contamination; that they spontaneously assembled themselves at some point in the Solar System's earliest history by purely chemical means; or, most intriguingly, that they were themselves the products of animated collections of organic biological molecules – life.

Thus began the arduous task of characterising the chemical composition of the organics inside the carbonaceous chondrites. One by one, carbon-based molecules were carefully parsed from these extraordinary meteorites and identified based on their chemical properties in laboratories across the world. Some were found to be simple, consisting of only one or two carbon atoms with short twigs of adjoining oxygen and hydrogen. But others were exceedingly complicated: snaking long chains of carbon, decorated with molecular arms that twisted outwards like gnarly branches; rings of carbon trimmed with decorative molecular protrusions, jutting out at all angles along the circumference. Carbon was joined by plenty of hydrogen and oxygen, and smatterings of nitrogen, phosphorus, and sulphur. The complexity and diversity of these molecules was staggering.

One meteorite that yielded a particularly rich harvest of organic molecules was a meteorite named Murray, which belonged to a group of carbonaceous chondrites called the 'CM chondrites' (named after a meteorite called Mighei that fell in Ukraine in 1889). CM chondrites are renowned for being packed full of organic molecules. In the small hours of 20 September 1950 in Kentucky (USA), after falling as a fire visible across five US states and sending forth a boom that rattled windows over a 2,600-square-kilometre area, lumps of Murray thudded to the ground. One piece rammed through the roof of somebody's house and landed squarely on the floor below. (Nobody was hurt.) In total, Murray weighed just over twelve kilograms, making it one of the largest CM chondrites known. But whilst a few pieces of the stone were recovered immediately, a formal search party was not assembled for several weeks, by which time it had rained.

In 1962, researchers at Caltech were delving into the organic broth that braided the rocky fabric of Murray when they made an electrifying discovery. They identified an evocative class of organic molecules that are essential to life as we know it: amino acids. As the elementary units of proteins, amino acids fit together like pieces of a jigsaw puzzle to form the biomolecules from which we – and all other life on Earth – are constructed, earning them the title 'the building blocks of life'. They play a central role in the molecular machinery that underpins metabolism and the maintenance of other life-giving processes.

Here they were, the building blocks of life, adrift in the sea of organic molecules lacing a CM chondrite. Over the inter-vening half decade or so, more amino acids were detected in different carbonaceous chondrite groups, such as the CI chondrites (including pieces of Orgueil that had not been tampered with). Things were starting to look more complex than previously thought. But a problem remained: the amino acids were present in trace amounts – typically mere millionths of a gram per sugar-lump-sized piece of meteorite – and could easily be terrestrial contamination.

Detecting tiny quantities of anything in meteorites – organic or otherwise – is exceedingly difficult: one fights a constant battle against background levels of contamination. In a 1965 *Nature* paper playfully entitled 'Amino-acids on Hands',[2] chemists showed that the smidgen of oil left behind in a single human thumbprint could easily be the source of apparent extraterrestrial amino acids in carbonaceous chondrites. Minuscule amounts of contamination would be enough to account for the quantities of amino acids typically measured in a meteorite. A few fingerprints here and there is all it would take (not to mention an accidental fingerprint on the pipettes, glassware, and measuring cylinders used in the laboratory). It would not take much to contaminate a whole chunk of chondrite with amino acids and give the false impression that it housed the extraterrestrial building blocks of life. The author of 'Amino-acids on Hands' noted: 'What appears to be the pitter-patter of heavenly feet is probably instead the print of an earthly thumb.'

Doubts about the origin of the amino acids were dispelled when a particular meteorite fell out of the sky in 1969. After sending shockwaves across the Australian outback and leaving a blue column of smoke in its wake, answers to the puzzle rained across the farmland surrounding the village of Murchison.

Handedness

Whilst the occasional 'print of an earthly thumb' would no doubt have been imparted onto most of its pieces, the Murchison meteorite was so huge that the amount of organic contamin-ation passed onto the stone would have been negligible in comparison to the amount of organics within it. Carbon-based chemistry in meteorites is usually explored in tiny crumb-sized pieces; Murchison provided an opportunity to delve into huge chunks.

Besides, the stench emanating from the stone immediately after its fall proved that at least some of the organics were defin-itely extraterrestrial.

Within one week of receiving pieces of Murchison, Professor Lovering at Melbourne University had classified the meteorite as a

CM chondrite.* Whilst not as water- or organic-rich as the rarer CI chondrites like Orgueil, the samples were incredibly fresh. It presented the perfect chance to tease apart the extraterrestrial organic molecules – especially the amino acids – on a meteorite unspoiled by the Earth's biosphere and a century of careless handling.

News of the Murchison organics soon reached the press, and just two weeks after the fall, the *Canberra Times* carried the sensational headline, 'Organic fossils in rare meteors'. It was, of course, an exaggeration (no such fossils had been discovered), but it certainly captured the mood at the time. As sheer chance had it, Murchison made Earthfall on the same day *Apollo 11* Moon samples arrived in Australia for scientific scrutiny and so stoked a public consciousness already hyped with the notion of space exploration. The happy coincidence was not lost on anybody, including Lovering, who declared: 'The Moon samples cost man $2.8 million an ounce. It was just lucky that we have got this meteorite for free.'

Being a little more careful than the mainstream media, the scientific world waited some fourteen months after Murchison's fall to declare what it had found. After rigorously employing the checks and balances that accompany all sound scientific enquiries, a research team spanning three American institutions (NASA's

---

* Lonewolf Nunataks 94101 – the meteorite I studied as a twenty-two-year-old intern at NASA's Johnson Space Center – was a CM chondrite. Before it was discovered by Antarctic explorers in 1994 and returned to the Johnson Space Center for careful curation, it had spent some few thousand years encased in the sterility of the East Antarctic Ice Sheet. I had to slice it open using a thin-bladed, diamond-encrusted saw that took the best part of a day. Eventually, the apple-sized stone lay cleanly cut into two pieces. The first thing I did was hold the freshly sliced surface under my nose and breathe a deep breath. The smell hit me like a sharp stick. Such a stench would not have been out of place in a chemistry classroom, and I remember describing it at the time as a cross between a petrol station and a damp towel. It was *musty*.

Exobiology Division in California, the Department of Geology at the University of California, and the Center for Meteorite Studies at Arizona State University) published their findings. What they found settled the debate about the amino acids for good.

The team analysed an entire ten-gram piece of Murchison (about the size of a peach stone), which for a meteorite is absolutely enormous. Their piece was from the core of a larger chunk, too, which reduced the risk of accidentally sampling any prints of 'earthly thumbs' imparted upon the meteorite's surface. The lump was pulverised and treated with a series of chemical deluges in a bid to draw organics from the stone. Thrice-distilled water was used throughout the procedure to reduce the risk of accidentally mixing Earthly amino acids – present in the 'background' natural environment – into the sample.

Seven types of amino acids were detected in the organic medley extracted from Murchison. Although present in exceedingly small quantities – mere millionths of one gram – they were still definitely *there*. Two of the amino acids in particular, 'sarcosine' and '2-methylalanine', immediately raised a few eyebrows, because they are not usually found on Earth in biological systems. It is unlikely these amino acids arrived in the stone as contaminants. This hinted strongly at an otherworldly origin.

The smoking gun that proved the amino acid's extraterrestrial origin was a peculiar chemical property: 'chirality', also known as 'handedness' (the word chirality is derived from the Ancient Greek χείρ (*kh-eir*) meaning 'hand'). Here is how it works.

It is impossible to superimpose your hands. Your right hand – though it is at first glance identical to your left – is fundamentally different, and for one simple reason: your hands are *mirror images* of each other. This is chirality. Feet have chirality, too, which is best exemplified by putting your shoe on the wrong foot.

Many organic molecules exhibit the property of chirality. As with hands, it is impossible to superimpose two molecules of different chiralities because they are perfect mirror images of each other. They overlap face to face as if reflected in a mirror, but never in the same orientation. Amino acids behave in this way: each has a doppelgänger – an identical chemical twin – with identical chemistry but the opposite handedness.

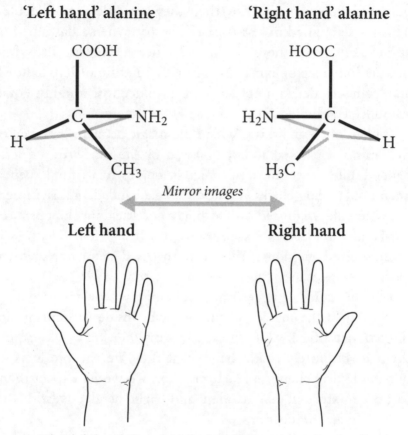

The 'left-handed' and 'right-handed' versions of a simple amino acid – alanine – found in Murchison in 1970. Whilst both versions are identical as far as their constituent atoms go – and the arrangement of atoms surrounding the central carbon – they are mirror images of one another. They cannot be superimposed, no matter how much they are twisted or turned.

This means that each type of amino acid, of which there are hundreds, can exist as one of two flavours – a 'left-handed' version and a 'right-handed' version – which are, aside from their opposing handedness, identical. If two amino acids contain precisely the same combinations of constituent chemical elements and feature the same groups of chemical appendages, but have different chiralities, then they are different molecules. This is exemplified in test-tube chemistry: when a flask load of amino acids are synthesised by purely chemical means (i.e. without the influence of living organisms), a mixture of 'left-handed' and 'right-handed' amino acids are produced in roughly equal amounts. But biological amino acids are different.

It turns out – and this is where the magic happens – that all of the amino acids used and synthesised by life on Earth are left-handed. Life as we know it, when it comes to its fundamental chemical building blocks, is entirely single handed. It is asymmetric. When life originated on Earth, it based itself on a blueprint of left-handed molecules. Exactly why it 'chose' left-handedness is an unresolved question. Perhaps carbon-based lifeforms elsewhere in the cosmos, if they exist, are right-handed.

But this quirk of terrestrial biochemistry provides us with a useful test. If the medley of amino acids inside Murchison are the products of living biological lifeforms, then we would expect them to be purely single handed; if they are the products of non-biological chemical reactions, then we would expect them to be a mixture of 'left-handed' and 'right-handed' versions in roughly equal proportions.

The cosmochemists discovered that the amino acids in Murchison are a two-handed mixture, comprising roughly fifty per cent 'left-handed' and fifty per cent 'right-handed' doppelgängers. This simple observation settled the case: these amino acids are not the products of life, Earthly or otherworldly. Whilst this (disappointingly) rules out a biological origin of the amino acids and other organic molecules in CM chondrites, it (convincingly) proves they are extraterrestrial.

Since its fall in 1969, tens of thousands of unique organic molecules have been identified in Murchison alone, but the number of molecules yet to be discovered most probably extends into the many millions. It's not just Murchison, either. An opulent mixture of organic molecules has been characterised in every single one of the eight groups of carbonaceous chondrites, as well as dozens of unique carbonaceous chondrites that do not readily fit into a group. Amongst them are more than seventy types of amino acids alone.

Across the Solar System

The abundance of organic molecules in the carbonaceous chondrites tells us that complex carbon-based chemistry was a widespread feature throughout the infant Solar System and that it was a common component of asteroids. Indeed, since we took to the skies and explored other worlds, we have discovered that many of them host similarly complex chemistry. Organic molecules seem to be the rule for celestial bodies rather than the exception.

Titan, the largest moon in the Saturnian system, is shrouded in a thick atmosphere of organic molecules, so much so that it rains liquid methane (CH_4). When the droplets fall upon its icy surface, they drain into lakes and seas of liquid hydrocarbons, carving deep networks of rivers and streams. Ultraviolet light emanating from the distant Sun initiates chemical reactions in Titan's high-altitude clouds, synthesising a cocktail of organic molecules probably as complex as those found in carbonaceous chondrites.

Organics have also been detected in the atmospheres of the gas planets – Jupiter, Saturn, Uranus, and Neptune – and on more of their many moons. In addition to Titan, organics have been detected on the Saturnian moons Enceladus and Iapetus. Three of Jupiter's giant moons – Europa, Ganymede, and Callisto – also house organic molecules, as does Neptune's giant moon Triton.

It's not just moons, either. In 2015, NASA's *New Horizons* spacecraft arrived at the Plutonian system after a nine-year voyage to the outer fringes of the Solar System. Pluto, a miniature world that lies some five billion kilometres from the Sun, is comprised of rock and a mishmash of ices: ammonia (NH_3), methane (CH_4), and water amongst them. Cosmic rays – originating from the Sun, distant stars, and other galaxies – irradiate these simple molecules and coax complex carbon-based chemistry into existence. These synthesised organics drape the frozen landscape, giving rise to Pluto's colourful patchwork surface.

After hurtling past Pluto at almost fifteen kilometres per second, *New Horizons* continued onwards for a further 1.5 billion kilometres (that's about the distance from the Sun to Saturn), where it encountered another world: 2014 MU_{69} Ultima Thule (formally renamed 486958 Arrokoth in 2019). It remains the most distant world ever visited by our species. *New Horizons* arrived on New Year's Day 2019 (what a way to start a new year) and streamed images of the city-sized world back to Earth over the intervening days and weeks.

The images are beautiful. They reveal a lumpy, peanut-shaped object, forged by the soft merging of two once-separate bodies, which like Pluto are made from a mixture of rock and ice. One of the most striking things about Ultima Thule is its colour. It is, like some patches on Pluto's surface, deep red. Practically its entire surface is covered by a layer of organic matter. It coats the miniature world like a layer of treacle, but lying almost seven billion kilometres from the Sun, these organics are frozen solid; I imagine they would feel like gritty snow underfoot.

At the turn of the twentieth century, ground-based astronomers discovered a small concoction of organic molecules in the
sweeping tail of Halley's Comet, spewing from its hissing surface
as it was boiled away by the heat of the Sun. One organic
molecule in the chemical medley – cyanogen (C_2N_2) – is infam-
ous for its toxicity. On 8 February 1910, the *New York Times*
emblazoned the words 'COMET'S POISONOUS TAIL' across
its front page. The story underneath hysterically reported that
Halley's tail could spell the end of life on Earth: 'Cyanogen is a
very deadly poison ... [and] would impregnate the atmosphere
and possibly snuff out all life on the planet.'

Enterprising businesspeople cashed in on the panic by flog-
ging gas masks and 'anti-comet pills' that were 'an elixir for
escaping the wrath of the Heavens'. Concerned citizens across
the United States even went so far as to seal their homes and
plug keyholes to ward off the deadly vapours. We can be thank-
ful that their efforts were in vain, and the whimsical predictions
of a planetary-scale apocalypse did not come to fruition. The
wisps of organics emanating from the surface of Halley into
space are so thin that they are barely there at all.

Organic molecules have even been detected in the ocean of
space between the stars in frigid nebulae, too, coaxed into existence by the cosmic rays streaming through the galaxy. It is an
arduous process that requires millions of years – luckily, in
cosmochemistry, we have billions of years to work with. Simple
free-floating gas molecules – such as carbon monoxide (CO)
and hydrogen (H_2) – are occasionally nudged by a cosmic ray,
which lends them the energy required to react with other gases
to form something more complex. The process repeats itself,
each time building progressively more massive and complex
molecules. These molecules are entirely out of reach and are
only detectable via detection of their spectroscopic fingerprints
using powerful telescopes.

And, of course, organic molecules are a feature of planet Earth, too, and they have taken many wonderful forms, including trees, cats, and humans.

Organic chemistry crops up on worlds in all corners of the Solar System. Carl Sagan, the widely celebrated astronomer, author, and science populariser, and his colleague Bishun Khare at Cornell University coined a term to describe these collections of molecules in a 1979 *Nature* publication.[3] Their name invokes a sense of what they would feel like to the touch in large quantities – something approximating gummy oil, maybe: 'We propose, as a model-free descriptive term, "tholins", although we were tempted by the phrase "star-tar".'

'Tholin' is from the Greek word θὸλος (*tholos*) meaning 'muddy', and while I do love it, I prefer 'star-tar'.

Origins

Sagan and Khare coined the phrase star-tar for good reason: some of the organic molecules in carbonaceous chondrites were inherited from the interstellar nebula that collapsed to form our Solar System. They are the surviving remnants – albeit slightly modified by the turmoils of Solar System formation – that were locked in a deep freeze after they were synthesised by cosmic rays. The interstellar organic molecules rained down onto the flattened plane of the protoplanetary disc, and swirled alongside some of the gas and dust that coalesced to form the first worlds.

Energy liberated by the collapsing cloud – and crucially, the light streaming from the new-born Sun at the centre of the disc – initiated chemical reactions between simple gas molecules. New complex molecules were kindled from simple ingredients. And the pre-existing interstellar organics were modified slightly: parts were stripped away and new groups of atoms bolted on in their place. Far from the inferno of the newly ignited Sun, the organics persisted. They circled the Sun in the outer fringes along with the swirling gas, motes of rocky dust, and hordes of ices in the protoplanetary disc.

After coalescing alongside the cosmic sediments to form carbon-laced planetesimals, the organics underwent further change. The energy liberated from the decay of lively radioisotopes warmed the parental asteroids of the carbonaceous chondrites. Watery concoctions of organics frothed. New molecular species were coaxed into existence on these asteroids, but the minute size of these worlds ensured they cooled quickly, stopping syntheses soon after they began. Locked inside the asteroids, these organics persisted for some 4.6 billion years.

The molecules of life are not difficult to make. Prebiotic organic chemistry is seemingly easy and happened at practically every stage of Solar System formation (and before). It is 'just chemistry', after all, and if Nature is sublime enough to forge systems of planets from mere wisps of nebula, She is certainly capable of assembling complex molecules around a skeleton of carbon, some of which, in time, would attain consciousness.

Watery worlds

Meteoritic water has been the subject of intense scrutiny, too, because practically every single one of the ~ 2,500 known carbonaceous chondrites (and many ordinary chondrites, too, for that matter) is laced with hydrated (water-bearing) minerals. Their presence proves that they were burdened with water before they arrived on Earth. The water bound up in the carbonaceous chondrites is indigenous to meteorites. It came from outer space.

Some meteorites, of which the CI chondrites such as Orgueil are the best examples, are comprised entirely of waterlogged minerals. These ultra-rare meteorites hail from asteroids that never melted, but, unlike most other chondrites, they do not preserve the dust from which their parent asteroids were forged. All of the cosmic sediment that coalesced to form the parental world of the CI chondrites was ruined by the action of water. All of the CAIs, chondrules, and matrix were utterly consumed, and were replaced by a suite of new, water-bearing hydrous minerals which locked the water away as they crystallised. This extraterrestrial water became bound inside their crystalline structures.

Many carbonaceous chondrites, like natural cosmic thermometers, recorded the temperature of the water, too. By characterising the assemblage of hydrated minerals that lace them (and their isotopic compositions), cosmochemists can read the temperature of the circulating water. Different meteorites recorded water at different temperatures, spanning from just-about lukewarm to borderline steam, but most of it was around the same temperature as bathwater.

The circulation of warm water – and the melting of the ice that originally coalesced alongside the motes of rocky dust – was not at all powered by the heat of the Sun. It was powered by the decay of short-lived radioactive isotopes, entirely independent of a stellar heat source. Far from being geologically dead worlds, the parent asteroids of the carbonaceous chondrites were dancing with hydrothermal activity, which in turn fuelled the synthesis of a rich concoction of organic star-tar.

In studying meteorites we plumb the depths of our origins, from the synthesis of the chemical elements to the formation of our Solar System and the assembly of rocky worlds like the Earth. The carbonaceous chondrites give this endeavour a biological slant.

There is no convincing evidence within any meteorite that it once contained anything alive. It may be that we just have not yet stumbled upon the signs of extraterrestrial life in a meteorite, but I doubt it. A far more likely scenario, to my mind (and the minds of many other cosmochemists), is that the carbonaceous chondrites rather record the chemical *prequel* to life's story.

They also directly point to the possibility of life elsewhere in the Solar System. Star-tar coats the surface and laces the rocky fabric of many worlds in our own planetary system, including many asteroids: the complex molecules of life are widespread and abundant, making it at least possible that they achieved animation elsewhere in the Solar System. Life originated on the Earth shortly after it had finished forming, after all, so what is to say the same did not happen elsewhere if the ingredients were available? There were countless asteroids in the nascent Solar System that housed warm, organic-laden water, which makes them the first potentially habitable places in the Solar System.

There is probably nothing inherently special about our own Solar System, either, and star-tar is undoubtedly a feature of planetary systems across the Universe. Even if organic molecules did not achieve animation elsewhere in our cosmic vicinity, the ubiquity of organic molecules in budding solar systems across the galaxy makes the emergence of life elsewhere in the Universe no less likely (and to my mind, inevitable). The chemical building blocks of life are everywhere.

Life inside asteroids, if it ever got going at all, would probably have been short lived, however. Even modestly sized carbonaceous asteroids would have cooled rapidly, chilling from the outside in as they lost their internal heat to space. Their sphere of habitability would have shrunk with the passing millennia, until eventually only their centres were warm enough. In time they froze throughout, and any lifeforms still clinging to the edge of existence would have perished in the cold. Perhaps one day we will send our spacecraft to these worlds – or go there in person – and sample the rocks beneath the surface, and find the frozen remnants of the Solar System's first forms of life.

Tantalisingly, we know for sure that, at least to a certain degree, primitive organics escaped their parental asteroids and landed upon the early Earth. Carbonaceous chondrites may be the seeds that were sown onto the surface of the early Earth from which the tree of life grew.

Earth's rock record recounts life's story; when we trace its history backwards along the branches through deep time, we look to the layers of sedimentary rocks that contain fossils: from the preserved remains of our ape-like ancestors, to the inhabitants of ancient oceans and the microbial mats that basked in shallow seas 3.5 billion years ago. If we follow this story backwards from the biological to the biochemical to the purely chemical, we eventually come upon stones from the sky.

Carbonaceous chondrites, in all their splendour, could well be our prebiotic ancestors from the skies.

9

PIECES OF THE RED PLANET

There are astronomical objects gracing the night sky that are visible in even the most light-polluted regions of our planet. One hangs in the sky in magnificent distinction as a giant silvery orb, so bright it can even be seen by daylight. It is, of course, the Moon. As a source of reflected sunlight gleaming brilliantly overhead, the Lunar surface floods night-time landscapes with silver and has inspired stories and mythologies for millennia. It is also a useful timepiece. Moonrise and moonset; wax and wane. The motion of the Moon across the night sky tracks the hours left until the Sun returns once more, and the monthly cycle of the Moon's shifting phases – from new, to crescent, to half, to gibbous, to full, then back again – marks the passage of around thirty days. The Moon was a convenient keeper of time for hunter-gatherers who relied on tracking the dissolution of one season into the next.

Other lights in the sky look like stars at first glance, but, when observed with close attention, reveal themselves to be distinct. These lights do not obviously wax and wane or appear as bright orbs like the Moon. If anything, they appear as normal stars, but with a few quirks and peculiar characteristics.

All stars seem to move across the sky as the night presses on; the great arcs they trace are a consequence of the spinning motion of the Earth, but their positions relative to one another do not change. It is as though they are fixed points of light shining inwards from an infinitely large dome that rotates over-head. Some 'stars' do shift their positions, though, and by an amount noticeable on timescales as short as days and weeks. They too trace great arcs across the sky as the night presses on, but they shift relative to the backdrop of fixed stars from night to night. These slowly moving points in the sky did not go unnoticed by the ancients. The Ancient Greeks called these peculiar lights πλάνητες ἀστέρες (*planetes asteres*) meaning 'wandering stars'. Today we call them planets, and they move across the sky because they, like the Earth, are orbiting the Sun.

For 200 millennia we humans were confined squarely to the surface of the Earth, both feet on the ground. Our stories were of Mercury the Messenger darting across the sky in great haste; of Venus, one of the most beautiful sights in the skies of Earth, aptly named after the Roman Goddess of Love and Beauty; and the angry-looking Red Planet, Mars, named in honour of the Roman God of War. The largest planet in our Solar System, Jupiter, is named after the King of the Roman Gods. And Saturn, the star that wanders at the slowest pace, takes its name from the Roman God of Agriculture who ruled the Earth during the 'Golden Age' of plentiful food and bounty. We personified the planets and, as we did with the stars, projected our own human characters and dramas upon them.

Little did we know that each had a story of their own to tell, some of which were written, like the story of the Earth, in rocks. By using the tools of science, we are unpicking their geological histories just as we did the Earth's. Within the space of a single human lifetime, spacecraft have been placed carefully into orbit around each of the six planets visible to the naked eye,[1] and we have safely landed robotic science laboratories on the surface of Venus and Mars. The exploration of our Solar System exemplifies the extraordinary technological feats that come about from dedicated and sustained effort, and show we are capable of solving even the most difficult problems, should we choose to bend our collective will towards them.

Meteorites, too, have played a pivotal role in our exploration of at least one of the wandering stars.

To Mars

On 27 November 1971, the Soviet Union's *Mars 2* space probe plunged into the Martian atmosphere at six kilometres per second, having departed Earth aboard a rocket just over six months earlier. Stowed inside was a portable science laboratory, designed to land softly on the Martian surface. A heat shield protected the precious lander, which was equipped with a Swiss army-knife array of scientific instruments, taking the brunt of the intense pressure and 1,000 °C temperatures during descent. An onboard parachute was tucked away inside, ready to carry the explorer safely to the surface. But the parachute failed. Three minutes after entering the Martian atmosphere, *Mars 2* fell silent. Despite being considered an engineering failure, *Mars 2* became the first human-made object to land on the surface of the Red Planet.

An identical spacecraft named *Mars 3* was launched just nine days later. Hot on the heels of its interplanetary twin, it arrived at the Red Planet on 5 December 1971, but was to suffer a similar fate.

The arrivals of both crafts were badly timed. A huge dust storm shrouded the entire planet as 100-kilometre-per-hour winds kicked clouds of sand and fine dust seventy kilometres into the sky. That's eight times the height of Mount Everest. Even Olympus Mons, the tallest volcano in the Solar System, was submerged. It remains to this day the most epic dust storm ever observed on Mars. The spacecraft carrying the landers were not carrying enough fuel to enter orbit with the landers still attached, and so it was impossible to wait the dust storm out. The Soviet scientists and engineers had no choice but to send the precious landers to their doom. They could do nothing to save them.

Mars 3 had an identical landing system to *Mars 2*. Miraculously, the computer-controlled landing procedure was executed perfectly, but the howling Martian gusts were enough to render the parachute almost useless. *Mars 3* crash-landed, and was battered beyond use. Twenty seconds after landing, *Mars 3* fell silent, but not before a faint hint of a signal reached Earth. It relayed a stream of data from the onboard tape recorder intended to capture the first photograph ever taken of the Martian surface. No photograph appeared; only a formless white shape was visible through the fuzz of pixels. *Mars 3*, like its twin, was never heard from again.

Success yet again eluded the Soviet Union in March 1974 when the *Mars 6* and *7* twins failed to land (softly) on the Martian surface. No electricity flowed through their intricate electronics; no instruments sat waiting to capture images and turn the Martian stones; no onboard circuit listened for a wake-up call from Earth. Each heap of metal and wires lay freezing cold and lifeless.

NASA were not to suffer the same misfortune. Learning from the mistakes of the doomed *Mars* landers, NASA assembled their own interplanetary space probes, launching a pair towards the Red Planet in August 1975: *Viking 1* and *Viking 2*. Each *Viking* comprised two parts – an orbiter and a lander.

Viking 1 arrived at Mars in June 1976 and slipped into an orbit around the planet. Within one month a landing site had been finalised and the lander was released. Nine minutes after entering the Martian atmosphere, *Viking 1* touched down softly atop firm red soil. We were finally there. Twenty-five seconds after landing, a photograph was beamed towards Earth. The event was broadcast live across the United States, and line scan by line scan, the first photograph taken on the Martian surface materialised on television screens.

It was, without doubt, one of the most important photographs even taken. It marked the moment that Mars, which for almost all of human history was little more than a red star wandering across the sky, became a place. One of *Viking 1*'s footpads casts a short shadow onto the rock-littered Martian soil. In the absence of human explorers, *Viking 1* was there on our behalf, and it was seeing for us through mechanical eyes.

Daily television specials broadcast the events unfolding on the Martian surface into living rooms across the United States. *Viking 1* issued the first ever Martian weather report; lows of −86 °C, an afternoon high of −33 °C, with strong gusts up to thirty-two miles per hour. Something as ordinary and Earthly as weather was now a reality on another planet. Within weeks *Viking 2* had arrived in orbit and soon joined its twin on the Martian surface.

Photograph after photograph was taken by each lander. Rocky cobbles, against which drifts of wind-blown sands piled, littered the ground around the *Viking*'s footpads; fields of rolling sand dunes draped the landscape; a crisp frost, made from a mixture of frozen carbon dioxide and water ice, clung to the sands and rocks during the winter months, coating the otherwise red landscape with a white finish. Mars had a horizon, and above it stretched a sky tinted orange by fine particles of wind-blown dust illuminated by dim sunlight.

Thirty days after landing, *Viking 1* photographed the Sun setting below the Martian horizon. After watching millions of sunsets on Earth, we are among the first humans to witness one on another planet.

Not of this world, or that world

In 2005, after 324 Martian days roving the red surface, NASA's *Opportunity* rover came across a particularly strange-looking stone. In stark contrast to the rugged orange stones common on Mars, this one had a polished exterior and was covered by furrow-like depressions. The furrows were oddly reminiscent of regmaglypts* found on meteorites here on Earth, and chemical analysis revealed it to be made almost entirely of iron and nickel. It was a piece of metal that had fallen from the Martian sky. It was an iron meteorite.

This piece of asteroid shrapnel was, following convention for meteorites discovered here on Earth, nicknamed Heat Shield Rock after the place it was discovered: next to the wrecked heat shield that had protected *Opportunity* as it plummeted to the Martian surface almost one year before. Because the Red Planet lacks the protection of a thick atmosphere, meteorites make it to the Martian surface with ease and do not readily rust once they land; falling on a barren, planet-wide desert, meteorites also do not suffer the destructive effect of rain. Nine more meteorites on the surface of Mars have since been discovered by the rovers.

Meteorites, we have found, are not exclusive to Earth.

In the late 1970s it was still widely held that all meteorites came from asteroids. Allan Hills 81005 – the first meteorite recognised as a piece of the Lunar surface – was not discovered on the frozen East Antarctic Ice Sheet until 1982. That a stone could be blasted from the surface of a large body, such as the Moon or a planet, without being entirely obliterated in the process, was not

* See page 139.

thought to be possible. Surely, went the rationale, stones could not survive the acceleration required to escape such strong gravitational fields.

Meanwhile, cosmochemists were well on the way towards organising meteorites into major groups based on their geological similarities, under the working hypothesis that each originated from a tiny world in the Asteroid Belt. All meteorites, went the reasoning, must be the shrapnel from asteroids. But occasionally, there was a meteorite that stubbornly refused to cooperate with this asteroidal worldview.

As the geological character and stories written within meteorites were unpicked on a deeper and more fundamental level, unexplained observations came to light: there was something amiss. Some meteorites showed signs that they originated from a celestial body far bigger than any world in the Asteroid Belt. Three such groups of meteorites in particular stood out as particularly odd.

Three strange stones

A meteorite belonging to one of these odd groups disturbed the lush autumnal French vineyards of Champagne-Ardenne in 1815. As it plummeted through the atmosphere, it rattled like a musket. No fireball was glimpsed in the cloudless morning sky but, alerted by the clanging sounds, a vintner watched a solid object fall from the sky and make landfall nearby. He approached the newly formed hole in the ground and found a stone. News of the fall spread throughout the local community. Folk from the nearby village of Chassigny quickly descended and discovered more fragments of the celestial stone – each piece coated by a black crust – taking the total mass of this strange stone to four kilograms. As per the tradition of naming meteorites after the place where they fall, this one was named 'Chassigny'.

A second odd meteorite fell in north-east India on a clear August morning in 1865 and, causing something of a sensation, was featured in the *Calcutta Gazette* which announced: 'A stone fell from the heavens accompanied by a very loud report, and buried itself in the earth knee-deep.' Farmers tending their fields saw precisely where the stone had made landfall and, like the French vintners fifty years before, they rushed to retrieve it. Within minutes of entering the Earth's upper atmosphere the stone was being turned in the hands of curious onlookers, and the five-kilogram meteorite was named 'Shergotty' after the village in which it fell.

On 28 June 1911, a shower of stones rained over the village of El Nakhla el Baharia in Egypt, forty-five kilometres east of the ancient city of Alexandria. A towering column of white smoke hung in the stone's wake and terrible explosions rippled across villages dotting the Nile Delta. Though initially frightened by the event unfolding in the skies, local residents rapidly began collecting pieces of the fallen stone – some fragments embedded an arm's length into the ground – and eventually over forty were recovered, weighing in at over ten kilograms. Most of the stones were encased in a particularly brilliant black fusion crust which only added to the intrigue. This meteorite came to be known as 'Nakhla'.*

Despite falling across three different continents over the course of around one century, the stories accompanying the falls of Shergotty, Nakhla, and Chassigny are remarkably similar: onlookers

---

* There is also a story, which has become something of a legend among cosmochemists, of a dog being hit and killed by a piece of Nakhla as it fell. It is impossible to verify whether or not this unfortunate story is true (I hope it is not).

were initially startled or even frightened, but their fear quickly melted away into curiosity and excitement. And while each of the three meteorites were different from one another, they shared an obvious similarity: each was an igneous stone that formed from the crystallisation of once-molten rock. When viewed in thin section down the eyepiece of a petrologic microscope, each group erupts into a kaleidoscope of interlocking igneous minerals, glowing with beautiful colours. They were also chemically similar. As more meteorites made their way into the hands of the cosmochemistry community over the following century, more stones similar to Shergotty, Nakhla, and Chassigny were discovered. Thus, three new groups, bundled together on the basis of geological similarity, were added to the system of meteorite classification. Being the first of their kind to be discovered, Shergotty, Nakhla, and Chassigny became their namesakes: the 'shergottites', the 'nakhlites', and the 'chassignites'.

It was clear that these three groups were somehow linked, and most cosmochemists strongly suspected that they originated from the same parent celestial body. Moreover, there was growing evidence that these stones originated from a world that was far larger than any known asteroid. The three separate groups of meteorite were therefore bundled into what became known as the 'SNC clan'.

Shergottites, nakhlites, and chassignites

The shergottites formed after magma oozed across the surface of a celestial body as a thick carpet of molten rock. As the glowing lava cooled, it crystallised to form an igneous rock by now familiar on Earth, the Moon, and some asteroids: basalt. When viewed side by side under a microscope, many shergottites are practically identical to the rocks comprising the lava flows of Hawaii and

Iceland. At some point during their history, Shergotty and the other shergottites were hammered by a high-energy impact – a small asteroid collision of some sort – on the surface of their parent world. Shockwaves from the impact shot through the basalt and left behind a criss-crossing network of cracks and molten veins that rapidly quenched to form glass.

Nakhlites are comprised almost entirely of large crystals of an igneous mineral named clinopyroxene. These crystals recount a tale of slow-cooling magma that, unlike those in the surficial shergottites, crystallised underground where it was insulated by overhead rock. It is likely that the nakhlites formed at the base of this thick band of underground magma: the large clinopyroxene crystals sank through the belt of magma and accumulated at its base, and as more clinopyroxene crystals rained from above, the pile grew thicker. Eventually the entire underground system of magma froze, trapping the crowded pile of accumulated crystals at its base, locking the clinopyroxene into place before they were subsequently disentombed by a large impact from space. Their exhumation must have been gentle, however, because they show practically no signs of having being shocked.

While the chassignites formed under similar circumstances, they are an entirely different type of igneous stone. We know of only three chassignites – the French vineyard fall and two African desert finds – making them an exceptionally rare and valuable type of meteorite. Chassigny, the only witnessed fall, did not suffer at the hands of the Earth's weather and oxygen-rich atmosphere, and so retains a pale peridot-green colour of its almost pure olivine mineralogy. The olivine from which it is made sank through an underground cavern of slowly cooling magma to form a pile of interlocking, green crystals. Upon excavation from their subsurface prison by impact, the chassignites were marred by intense pressure, and the trauma of the shockwaves warped and fractured some of the olivine.

But each group also contains traces of something not normally associated with stones crystallised from molten rock: water. Many SNC meteorites are laced with crystals that precipitated from briny, mineral-laden waters. As water circulated through these stones and percolated into fluid gaps between their igneous crystals, it left behind minerals familiar to geologists who specialise in studying rock that has been affected by water: carbonates, clays, and salts. This opened up the prospect that the SNC meteorites originated from a celestial body which housed not only powerful igneous activity, but also freely flowing water.

Delving a little deeper, cosmochemists at the University of Chicago measured the blend of oxygen isotopes in a handful of SNC meteorites. This was the same team who discovered the exotic oxygen isotope signature of the CAIs some ten years previously; their expertise was unparalleled and even by today's standards their data remain impressive. They discovered that all three groups of SNC meteorites were similarly lacking in the lightest isotope of oxygen, ^{16}O. This proved beyond reasonable doubt that the SNCs originated from the same celestial body.

More than that, when the isotopic blends were plotted on an oxygen isotope plot (like the one on page 47), the SNCs traced a single straight line. The line had steepness ½. It ran exactly parallel to the terrestrial fraction line – the line traced by all Earthly oxygen, which formed as oxygen was smeared out through chemical, geological, biological, and physical processes over geological time. The line traced by the SNC meteorites was named the 'SNC fractionation line'.

Oxygen had revealed a common parental world that was geologically evolved and complex. Wherever the SNCs originated, geological processes must have acted extensively, smudging out the oxygen to produce the line of slope ½. It was a world with a geological history far richer and more complex than the relatively simple igneous asteroids, which, having frozen within a few million years of melting, did not have the time to shuffle oxygen around to produce long lines of fractionation.

Isotopic insights

As is so often the case in cosmochemistry, it was isotopes that provided further insights. The antiquity of meteorites is one of their distinguishing features; only the asteroids, the tiny parent bodies of the meteorites, have the means to preserve rocks from the first few million years of our Solar System's history. The naturally occurring atomic clocks inside meteorites, steadily ticking as radioactive isotopes decay to form new elements over the passage of time, all record ages in the region of 4.6 billion years. The CAIs, the chondrules, the eucrites and diogenites, the iron meteorites, and the other groups of volcanic rocks from differentiated asteroids all crystallised during the Solar System's infancy.

The only obvious way a meteorite could yield a younger age would be if somehow its internal clock had been reset by the shock of a large impact. Such shocks tend to write obvious tales of destruction into a rock's fabric. It is rather puzzling, then, that some SNC meteorites seem to be geologically young despite showing no obvious signs of being perturbed by an impact.

Nakhla, for example, is a nearly pristine igneous rock, which crystallised from its parental magmas around 1.3 billion years ago. This is youthful by normal meteorite standards. The asteroids formed 4.6 billion years ago; some SNCs, tantalisingly, are three billion years younger. This also places an upper boundary on the time at which water was freely flowing on the parental world of the SNCs.

The shergottites, nakhlites, and chassignites came from somewhere in the Solar System – excluding the Earth – that experienced igneous and hydrothermal activity in the geologically recent past. A volcanically active world! With water! In one fell swoop, the asteroids were eliminated as a possible source, and the number of potential parent worlds for the SNC meteorites fell from over one million to just four: Mercury, Venus, Jupiter's moon Io, and Mars.

A process of elimination

Mercury's surface is comprised entirely of igneous rock similar to Earthly basalts. Swarms of volcanic vents have been photographed from orbit, and, in places, tongues of frozen lava have been discovered stretching across vast Mercurian planes. It is, in almost every respect, an igneous world. But Mercury is entirely mottled by craters – countless numbers of them – indicating that, like the Moon, it has been a geologically dead planet for billions of years. If Mercury had been volcanically active in the recent geological past, many of the craters would have been filled with lava, rendering them invisible to orbiting spacecraft. Mercury's surface is truly ancient, and its surficial basalts far too old to be the source of the relatively infantile SNC meteorites. Besides, any rock ejected from the surface of Mercury would most likely be captured by the intense gravitational field of the Sun and soon meet a fiery demise;* it is incredibly difficult to launch rocks from Mercury outwards into the Solar System and all the way to Earth.

This rules out Mercury, leaving only Venus, Io, and Mars.

Venus, like Mercury, is an igneous world with a surface made almost entirely of crystalline basalt; but unlike Mercury, Venus is

---

* Computer simulations have shown, however, that it *is* possible for rocks ejected from Mercury's surface by impacts to make it out of the Sun's deep gravitational well. Some of these may be swept up by Earth and fall as Mercurian meteorites. None have (yet?) been recognised as such, however.

largely void of impact craters, indicating that it was resurfaced by fresh lava in the relatively recent geological past, perhaps some one billion to half a billion years ago. This is roughly, though admittedly not perfectly, in line with the crystallisation ages of the SNC meteorites.

However, Venus' thick carbon dioxide atmosphere is so soupy that it prevents all but the most gigantic of impacts from reaching its surface. This protective blanket over the whole planet works both ways, though, and keeps rocks in as much as it keeps them out. Even rocks ejected from the Venusian surface by an impact at hypersonic velocity are slowed to a standstill before making it above the clouds. Blasting a rock from the surface of Venus is virtually impossible. Besides, the planet is entirely arid, making it difficult to explain how the water-formed minerals lacing the SNCs could have formed there.

This rules out Venus, leaving only Io and Mars.

In 1979, NASA's *Voyager 1* space probe discovered the most volcanically active celestial body in the Solar System: one of Jupiter's giant moons, Io. Nobody had predicted such a strange world could exist in orbit around a planet. Hundreds of volcanoes, over 150 of which are active today, spew forth rivers of molten rock and launch towering clouds of ash hundreds of kilometres into the Ionian sky. Io is a perpetual inferno. Continually re-coated with fresh blankets of igneous rock, Io's surface is completely free from impact craters because as soon as they form, they are filled by lava.

Io, at first glance, is a strong candidate for the SNC parent body. We have photographs of volcanic plumes billowing forth from its surface upwards into space, and it is not unreasonable to speculate that rocks may become entrained in the upwelling columns of ash and gas. If not the volcanoes, then an impact from even a small asteroid could eject rocks from its surface into space, unimpeded by an atmosphere. But any rock ejected from the surface of Io would immediately be dragged towards Jupiter by its enormous gravitational field and be consumed. Stray Ionian stones are confined squarely to the Jovian system. The final nail in the coffin is the moon's extreme dryness: it is the driest celestial body in the Solar System and contains absolutely no trace of water at its surface.

This rules out Io, leaving only one possibility.

Ideas and hypotheses were flowing quickly through the cosmochemical community. By the early 1980s, several research groups across the world had converged and hypothesised the seemingly impossible: the shergottites, the nakhlites, and the chassignites came from Mars.

Mars fitted the bill in every way. The igneous nature of the stones; the former presence of freely flowing liquid water (ancient river channels on Mars had been photographed from orbit by this point); the young crystallisation ages of all three groups of meteorite, indicating that they originated from a large, recently active world. Yet somehow, it still seemed outrageous.

Not everybody was convinced. Moreover, no meteorite had yet been discovered that originated from any celestial world besides an asteroid, but that all changed in 1982 when Allan Hills 81005 was unambiguously tied to the surface of the Moon by comparison with the *Apollo* samples. Allan Hills 81005 proved that stones could survive ejection from large parent bodies.

Admittedly, the evidence that the SNC meteorites came from Mars was, while compelling, only circumstantial. The stones were tied to the Martian surface by a string of reasoning one would expect from Sherlock Holmes rather than anything chemical or isotopic. After all, no definitively Martian stone had ever been chemically analysed in an Earthly laboratory for a side by side comparison. But Mars *had* been chemically analysed in Martian laboratories: those onboard the rovers.

Discovery on the ice

While the drama over the origin of the SNC meteorites was unfolding in temperate latitudes, Antarctic explorers at the bottom of the globe discovered a stone that would change everything. A meteorite, the size of a cooking apple, was picked up from atop the barren ice sheet. In parts, white–grey rock was visible from beneath a veil of black fusion crust, but on the face of it, it did not look particularly special. It was named after the stretch of the East Antarctic Ice Sheet on which it was found and the year that it first made its way into human hands: Elephant Moraine 79001. It was classified as a shergottite. Like the other shergottites, it was comprised of interlocking magmatic crystals and was laced with glass formed during a violent shock.

Recognising that this meteorite belonged to a special group, cosmochemists at the Johnson Space Center set themselves an ambitious goal: to liberate minute quantities of gas trapped inside the stone when it was shocked, and measure its precise chemical composition. They hoped that it could help shed some light on where exactly it came from. Their tool of choice was a furnace attached to a mass spectrometer.

Breaking off a small chip of Elephant Moraine 79001 that showed clear signs of shock, they placed it inside their miniature furnace and sucked the air out of the entire instrument using a small army of powerful pumps. Even tiny amounts of stray Earthly atmosphere left inside the spectrometer would contaminate the minute amounts of gas trapped inside the shergottite. This experiment had to be performed in an outer space-like vacuum.

Slowly they turned up the temperature in the miniature furnace, and as they did, gases were released from inside the rock. They chose to measure the blend of noble gases. The first breath of gas exhaled by the stone was air from the Earth's atmosphere that had been absorbed onto its surface. Upon being relinquished from the crystals, the trapped gas flowed along a series of sealed pipes and was funnelled into the spectrometer for precise chemical fingerprinting. As expected, it was a match for Earth's atmosphere. Nothing particularly exciting.

Once Earthly air had been driven from the stone by tepid temperatures, the cosmochemists turned up the heat. Slowly the rock began to soften, then it began to glow, then to melt. With each step upwards in temperature, microscopic bubbles inside the stone were unwrapped and popped, releasing minute puffs of trapped gas into the furnace before going into the spectrometer.

Testament to the stone's celestial origin, the gas trapped within its crystals contained a blend of noble gases unlike anything on planet Earth. But the blend of noble gases was similar to one measured seven years previously, not on Earth by cosmochemists, but on the surface of Mars by a robot.

Viking 1 had inhaled a mechanical lungful of Martian air and measured the cocktail of elements and isotopes in the atmosphere. The gas trapped inside Elephant Moraine 79001 had the same fingerprint as the *Viking* measurement. Elephant Moraine 79001 contained bubbles of trapped Martian atmosphere.

The conclusion was inescapable and astonishing. Elephant Moraine 79001, along with the other shergottites, was a piece of Mars. Nakhla and Chassigny, tied to the shergottites by their common oxygen isotope fingerprints, must also be pieces of Mars. Shergotty, Nakhla, and Chassigny, and all of the other SNC meteorites, were pieces of the Red Planet that had rained from the skies to Earth. They were pieces of Mars.

The shergottites, nakhlites, and chassignites were thus renamed the 'Martian meteorites'. They remain one of the only groups of meteorites for which we know the parent celestial body with any degree of certainty; the others are the Lunar meteorites from our own Moon. Even the HED meteorites, which are probably from the asteroid Vesta, are only linked to this particular asteroid by circumstantial evidence. That means that of the 60,000 individual meteorites known to science, there are only around 650 (~ 400 Lunar and ~ 250 Martian, at the time of writing) we have convincingly tied to a parental body. Matching the remaining 59,000 or so meteorites with celestial bodies in space – practically all of which are comparatively tiny asteroids – is one of the most complex and difficult challenges of cosmochemistry.

Shergottites, nakhlites, and chassignites did for Mars what the *Apollo* samples did for the Moon: they helped turn it into a place with a story. In falling to the Earth, the Martian meteorites connect us physically to the red wandering planet that our ancestors gazed upon wistfully for hundreds of millennia.

No mention of meteorites from Mars is complete without telling the story of one of the most controversial meteorites ever discovered: Allan Hills 84001. After a sixteen-million-year journey through interplanetary space, this rock fell to Earth in Antarctica and was picked up from atop the ice sheet in 1984. This meteorite immediately sparked excitement, and notes taken in the field upon its discovery describe a greyish-green achondrite accompanied by the words, '*yowza yowza*'. The rectangular stone was returned to the Johnson Space Center along with the rest of that year's meteorite haul for classification and curation.

Beneath the black fusion crust, half-centimetre crystals of orthopyroxene dominated the rocky fabric of the stone. This was an igneous rock that had cooled deep underground where the mosaic of interlocking crystals could grow to gargantuan sizes. Allan Hills 84001 was also Martian.

A unique piece of Mars

The geological character of Allan Hills 84001 was wholly unlike that of the shergottites, nakhlites, and chassignites, however. It was far more primitive, far less geologically evolved, than the other Martian meteorites. It represented a completely new type of Martian meteorite that crystallised deep inside the Martian crust. It was and still is unique.

Allan Hills 84001 was also two billion years older than any other Martian meteorite, with a crystallisation age of a stunning 4.1 billion years. That is a mere 500 million years after the formation of the Solar System, and billions of years older than any of the other Martian meteorites. This stone formed in an ancient volcanic system on the red planet.

After the massive crystals grew from their parental magma, the stone was afflicted by meteorites smashing into the Martian surface. Parts of its original igneous texture were demolished as minerals were shattered by intense pressure, and shockwaves crumbled parts of the stone into fractures packed with crushed crystals. Allan Hills 84001 was left partially wrecked and marred with cracks.

But Allan Hills 84001 had deeper secrets to divulge. It formed so long ago, went the logic, that it could hold clues as to what Mars was like in the canyons of the planet's deep geological past. There was strong evidence captured by orbiting spacecraft that Mars had freely flowing liquid water on its surface in the distant past – oceans, rivers, and deltas – and analysis of Allan Hills 84001 was a way of exploring these ancient environments further. Rather than forming on Mars as we see it today, this stone formed on Mars when it might have looked something like Earth: waterfalls rumbling through canyons; rivers flowing into ancient lakes; rock pools dotting shorelines.

Traces of Martian water are laced throughout Allan Hills 84001. Hydrothermal fluids, laden with dissolved minerals, circulated through cracks in the Martian crust, leaving behind veins of water-formed carbonates in the rocks through which they seeped, similar to kettles and shower heads getting crusted with limescale. Precipitated carbonates twined through the cracks in Allan Hills 84001 like grouting between tiles. While the exact temperature at which these carbonates precipitated is unknown, it is possible that they formed in temperatures no higher than bathwater. Two independent isotopic atomic clocks recorded the time at which the carbonates precipitated; within 100 million years of the initial crystallisation of the stone.

Water circulating through conduits in the rock; the prospect of comfortably warm underground temperatures; a planetary surface that may have resembled Earth. The fleet of Martian space probes and the Martian meteorites had opened up a real possibility: that conditions suitable for life could have existed in Mars' deep past.

Electrifying claims were soon published in the journal *Science* in 1996 by a group of researchers led by an astrobiologist named David McKay.[2] This publication has since gone on to be one of the most (in)famous in the history of cosmochemistry. McKay and his team of eight co-authors broke small chips from Allan Hills 84001 and imaged them at the Johnson Space Center with an array of powerful electron microscopes, revealing the structure of the stone on unimaginably small scales. Down in the realm of geology on the nanometre level, strange features came sharply into focus.

McKay and his team discovered smidgens of organic compounds smattering the inside of Allan Hills 84001. Their absence in the stone's fusion crust proved their Martian origin; if they were the product of an invasion of terrestrial lifeforms during the stone's 13,000-year residency on the East Antarctic Ice Sheet, they would have invaded the stone's fusion-crusted exterior. But they were only found in its interior. The organics were not randomly distributed through the interior of the stone, either, but only appeared in the hydrothermal veins. These complex molecules were in some ways chemically reminiscent of the carbonaceous star-tars.

Inside the hydrothermal veins and dotted among the organics, strangely shaped elongated minerals mottled the surfaces of some of the carbonates. To some, these innocuous globs resembled tiny egg-shaped blobs, or perhaps long pieces of segmented rice. To McKay and his team, they looked like something else entirely – the fossilised remains of long-dead microbes.

Indigenous organic molecules; microbe-shaped globs; a meteorite that formed when Mars had water flowing across its surface and through its rocks. Despite being no longer than a mere 100 nanometres in length, McKay and his team interpreted the herds of rice-shaped globs to be anything but an innocuous fleet of sculpted carbonates. They believed they really were looking at the stony remains of nanoscopic microbes, fossilised inside the carbonate veins of Allan Hills 84001 some four billion years ago when warm water circulated through Mars' crust. The organics, they argued, were the chemical remains of those microbes, trapped inside the stone. They claimed to have found the fossilised remnants of life in a Martian stone, and the first evidence for extraterrestrial life.

A presidential stone

The publication immediately caused a media sensation, and the possibility of Martian life exploded into the public consciousness. Headlines worldwide spoke of life on the Red Planet and images of the supposed Martian nanobacteria lavishly decorated the printed press and television screens. So big was the news that President Bill Clinton gave a press conference about it on the

South Lawn of the White House before an audience of journal-ists and television cameras. In the period after the *Apollo*-era and the Space Race heydays, public and political appetite for space exploration had somewhat waned, but Allan Hills 84001, a meteorite fit for a presidential address, was to reignite it.

Debate immediately ensued within the scientific community. Most scientists were (rightly) incredibly sceptical of such a claim, and before long, some of the best minds in cosmochemistry were focused on Allan Hills 84001. There is not a single idea in the history of science that has been accepted on the basis of one publication, especially ideas that have ramifications as enormous as finding evidence for life on Mars.

One thing that immediately raised strong doubts – something that was acknowledged by McKay and his team – was the size of the so-called fossil microbes. They are tiny, far smaller than almost any known lifeform here on Earth, fossil or living. Some argued (and continue to argue) that Martian life may work in vastly different ways to Earthly life; maybe cellular life on such a minute scale operates comfortably on Mars, and to expect it to look exactly like life on Earth is narrow-minded. Maybe. Even so, life on that scale is exceedingly uncommon on Earth and is practically absent in the fossil record, raising serious and legit-imate doubts about the nature of the 'nanofossils' in Allan Hills 84001. On a planet like Earth that harbours such an astonishing diversity of lifeforms, would one not expect to have found example after example of such minute forms of life?

Even after sitting on the Earth's surface for 13,000 years, part of Allan Hills 84001's story was yet to be written, and the mere act of being picked up by human hands became part of the meteorite's multi-billion-year history. We humans sometimes unintentionally change the subtle nature of rock in our labora-tories, and we can inadvertently write stories of our own into them. These human stories can be difficult to untwine from the non-human chapters. Within one year of the explosive paper being published in *Science*, cosmochemists in other laboratories had recreated the 'microfossils' in their own pieces of Allan Hills 84001. Some researchers found that in preparing their sample for imaging via electron microscopy – using identical preparation techniques to McKay and his team – they had accentuated natural features of the rock to make them look microbe-like. They even artificially recreated the segments in the elongated globs. It looked likely that the 'microfossils' in Allan Hills 84001 were indeed the product of life, but Earthly rather than Martian; specifically, they were a product of human imagination.

Even the presence of organic molecules in Martian meteorites are not a strong argument for the presence of life. Although the cyclical chains of carbon in Allan Hills 84001 is an exhilarating discovery and a marvel of analytical cosmochemistry, they are relatively simple molecules. They do not even approach the chemical sophistication of the star-tar in carbonaceous chondrites. And an enormous cavern of complexity separates even star-tar from the chemistry of bona fide life. All life that we know of requires complex organic chemistry, but not all complex organic chemistry requires life.

There is no evidence to rule out definitively a biological origin for the strange forms in Allan Hills 84001, but, as Carl Sagan said: 'extraordinary claims require extraordinary evidence'. Scepticism and exceptionally high standards of consistent, reproducible evidence lie at the heart of the scientific enterprise. All scientific hypotheses are put through a rigorous set of checks and balances. Ideas are debated freely in writing through journal papers and in person at annual conferences and meetings, and only those that can withstand the full scrutiny of the collective scientific mindset survive.

Claims of discovering fossil microbes in a Martian meteorite require the most extraordinarily strong evidence. No such evidence exists. The testimonials written into the geological fabric of Allan Hills 84001 are, at best, ambiguous, and most of the cosmochemical (and biological) community remain unconvinced of McKay and his team's conclusions.

Allan Hills 84001 is a cautionary reminder that sometimes we see what we hope to see. It is easy to fool ourselves. Nature, in all of Her wonderful complexity, forges many strangely shaped objects, and some of them resemble the remains of once-animated lifeforms even if they are entirely non-biological. Shape alone is not enough to identify fossil life on the microscopic scale.

On 31 January 2014, NASA's *Curiosity* rover captured a photograph of the Martian sky as night was drawing in. The Sun had set below the horizon eighty minutes before, and the dusk was fading to black behind rolling hills on the horizon. A blue point, outshining all other sources of light, gleamed like a beacon in the sky. If *Curiosity* were to watch this brilliant light night after night, it would notice it was not a fixed star at all, but instead slowly wandered across the sky against the fixed backdrop of stars. Our ancestors would have called it a wandering star; the blue beacon in the sky was a planet. It was Earth.

Mars has a sky and a horizon. Weather blows sand across its surface and crisps the landscape with cold frost. The Sun sets over its rolling dunes and mountains, casting long shadows over a landscape beneath which lies a planetary history written in rock, just like the history of the Earth. The robotic explorers we send forth to sift Mars' red soil and turn its rocks are our planetary ambassadors, testing the waters before we travel there ourselves.

We have come so far in such a short space of time on our journey towards becoming an interplanetary species, and now, with the advent of human space exploration, we are ready to journey back to the Moon and the planets beyond. We will settle on the surfaces of other worlds, at first temporarily, but eventually permanently. We will leave our footprints in new sands as we watch the Sun setting behind new horizons, and our bodies will cast long shadows over new and unmapped land-scapes. We will never forget where we came from – our precious blue ocean planet shining in the sky – but some of our descend-ants will one day call a place other than Earth 'home'. It is not impossible that some of those people have already been born and walk amongst us.

Turning the rocks of Mars beneath the red soil and reading the geological history written within, we will discover a tale every bit as beautiful as the one written here on Earth. We will, piece by piece, reconstruct the geological and climatic history of Mars, expanding our knowledge and understanding of the possible paths a planet can take through the course of deep time. We will scour the volcanic provinces of Mars in search of rocks that resemble the shergottites, nakhlites, and chassignites; even-tually we will discover the exact point on the surface from whence those meteorites came before embarking on their ten-million-year journey through interplanetary space to the Earth.

The memory of those who ventured there before us will go down in the Martian history books. The crumpled remains of the *Mars 2* and *3* landers will become sites of pilgrimage and symbols of perseverance, reminding us of those failed early attempts. The landing sites of the NASA rovers – the first Martian explorers – will become sacred sights held in as much reverence as the holiest of sites here on Earth; among them will be the *Viking* Memorial Monument, the *Opportunity* Visitor Centre, and the *Curiosity* Museum of Science and Martian

Exploration. The metal chassis of the long-dead rovers will take centre stage in each, preserved for all to see in their final resting places. Perhaps Heat Shield Rock will become the Martian Centre of Cosmochemistry (a part of me secretly hopes it will become a music venue).

Of the 60,000 celestial stones we have in our collection here on Earth, fewer than 300 of them are Martian in origin. That makes them some of the rarest, most valuable, scientifically important rocks known to humankind.

But they are much more than mere objects of scientific curiosity. Martian meteorites are (for now) the only way by which we can scrutinise the rocks of the Red Planet in scintillating detail. Thanks to them, and the fleet of Mars rovers, we have found the answers to questions that were first asked in antiquity about the bright point of red light that graces the night skies of Earth. There is still much to be learned about the fourth-nearest planet to the Sun, but the Martian meteorites have given us a tantalising glimpse of some of the secrets that lay undiscovered beneath the planet's surface. The Martian meteorites are, in a real sense, pieces of our future home.

CALAMITOUS TALES

Sixty-six million years ago, the Cretaceous Period was drawing to a close. The tranquillity of fern-carpeted forest floors, beneath swaying coniferous canopies, bore no signs of the turmoil that would soon arrive from the sky. After their 180-million-year tenure on planet Earth – just shy of one hour in our twenty-four-hour geological day – the long age of the dinosaurs was nearing its end.

An asteroid, twenty or so kilometres wide, came upon the Earth after an interplanetary voyage from the Asteroid Belt. The atmosphere did little to slow it down. Streaking groundwards with hypersonic velocity, it ploughed into the Earth's surface with no forewarning, and Armageddon ensued as hundreds of billions of atomic detonations worth of energy were released in an instant.

A seismic wave generated by the explosion pulsed through the surrounding landscape like a ripple emanating from a stone cast into a lake, tearing the ground into pieces as it went. Thousands of trillions upon trillions of tonnes of solid rock were gouged out of the ground and sent skyward as giant blocks, lobes of molten rock, and as angry snakes of gas. At the point of contact, temperatures soared to over 20,000 °C, some three times hotter than the visible surface of the Sun. Rocks which bore the full brunt of the impact were instantly turned into gas, utterly unmade on the nanoscale and reduced to their constituent atoms.

Energy from the explosion and the rapidly expanding gas blasted a curtain of stone outwards from the site of the impact and high into the air. The expanding wall of rock was travelling at fifteen times the speed of sound in all directions: at five kilometres away, you would have been hit by the barrage and pulverised a whole fourteen seconds before the deafening roar reached your ears. Boulders the size of buildings were nestled in and amongst the debris, and within seconds, billions of tonnes of rock began tumbling and blanketing the broken landscape.

A layer of jumbled wreckage – from the building-size boulders to pulverised motes of dust – covered the landscape in all directions, easily deep enough to bury most of the buildings in London. The air remained thick with debris for a while after. A gaping crater in the Earth's surface had formed and was filled with a glowing pool of molten rock mixed with fragments from the blast. It would have looked like a portal into the Underworld. As the larger stones plummeted from the sky like falling bombs, a towering ash cloud billowed up on high, but it too eventually tumbled groundwards. In the minutes, hours, and days following the impact, it settled like falling snow onto the devastation below. Still roasting hot by the time it landed, the baking layer of ash issued jets of angry gas through cracks and conduits, and, as it slowly cooled, the loose jumble was baked into solid stone.

Above the blast, the atmosphere was super-heated to incandescence and a firestorm raged across most of the planet. Earth's ecosystem was incinerated; forests were felled; tsunamis tore across the oceans. Soot and ash blocked sunlight from reaching Earth's surface for close to a year after the impact, inhibiting plants from effectively photosynthesising and thereby slashing the base of the food chain. Sulphurous particles injected into the atmosphere severely acidified the rain, which, after flowing into the oceans, acidified the oceans; the marine ecosystem was turned upside down as a result. Global temperatures dropped by at least several degrees Celsius, decimating entire ecosystems. Famously, the dinosaurs met their demise in the climate turmoil that followed. An event of such apocalyptic magnitude leaves a permanent trace of itself in Earth's rock record and reveals itself readily through the lens of modern science.

A line in the stone

In the sequence of rocks that straddle either side of the sixty-six-million-years-ago mark, there is a thin layer of sediment that sharply separates two distinct types of geology: below this single layer, the rocks contain dinosaur fossils, and above it, they do not. There in one layer; gone in the next. It is as though the dinosaurs disappeared in a geological instant of time. This layer of rock marks the point in time at which the dinosaurs suddenly became extinct and separates two great chapters in Earth's story: the Cretaceous Period (145 to sixty-six million years ago) and the Paleogene Period (sixty-six to twenty-three million years ago).

And it was not just the dinosaurs, either. Droves of marine species (such as the spiralled ammonites) were snuffed out, too, and plant life took an acute hit in what became a planetary-scale mass extinction. The knock to Earth's biodiversity was abrupt and severe, and remains one of the single biggest blows to the tree of life since its inception some four billion years ago. The mass dying takes its name from the two geological periods which it separates: the Cretaceous–Paleogene mass extinction event, or the K–Pg extinction for short.*

Whilst the global decimation of biodiversity is clear to read in the fossil record, for a long while the cause of the K–Pg extinction event remained a mystery. The answer, hiding in plain sight,

---

* *K* and *Pg* are the abbreviations for *Cretaceous* and *Paleogene*, respectively.

lay written in the rocks but it proved exceedingly difficult to decipher. Hypotheses ranged from the insidious (such as gradual changes in the global climate that pushed the environment over a tipping point), to the bizarre (such as a giant freshwater lake that suddenly drained into Earth's salty oceans, upturning the entire ecosystem), to the interstellar (a nearby supernova that bathed the Earth in deadly radiation). Yet physical evidence in support of any single hypothesis remained scant, and ideas remained, at best, speculative.

In 1980, however, the rocks began to tell their story. A team of four scientists from the University of California – one physicist (Luis Alvarez), one geologist (Alvarez' son, Walter), and two chemists (Helen Michel and Frank Azaro) – measured the chemical make-up of the thin layer of sediment that divides the Cretaceous rocks and the Paleogene rocks. The K–Pg boundary layer, went the logic, was deposited around the same time as the extinction event, and so it is probably the best place to search for the cause of the calamity. If any layer of rock were to contain an account of the turmoil, this would surely be it.

The team homed in on rocks from the face of a gorge in the Umbrian Apennine mountains of northern Italy and the sea cliffs of Stevns Klint near Copenhagen, Denmark. Both locales preserve a beautiful sequence of sedimentary layers spanning the Cretaceous Period and the Paleogene, and, crucially, the scientists were able to capture the layer of rock – only one-centimetre thick – that separates the two.

Within the rocks lay something strange. Back in their laboratory, the team discovered that the layer of sediment that separates the Cretaceous and the Paleogene is enriched in an uncommon chemical element named iridium. In a sugar-lump-sized piece of rock on Earth's surface, one might typically find just one billionth of one gram of iridium, but in the K–Pg boundary layer, it was enriched up to several hundred times. This is exceedingly rare. To double check their findings, the team also measured the abundance of iridium in rocks either side of the boundary layer. The strange iridium spike was confined squarely to the thin K–Pg boundary layer. Whatever caused the mass extinction had something to do with iridium.

A siderophile element, almost all of Earth's iridium was sucked into its metallic core with the sinking iron some four and a half billion years ago when it was a molten inferno. Only tiny quantities of iridium exist today in the rocky portions of our planet – the mantle and the crust – but there are rocks from elsewhere that contain it in great abundance: meteorites. Iridium was the first piece of physical evidence that hinted towards something of an extraterrestrial nature being involved in the K–Pg mass extinction event.

Alvarez and his team recognised the tantalising chemical connection between the K–Pg boundary layer and asteroid rocks (i.e. meteorites), and so they hypothesised that the mass extinction was caused by an enormous asteroid striking the Earth. They reasoned that if the asteroid was large enough, it would have delivered such a blow to Earth's surface that it would have upended the entire ecosystem, and the elements from which it was made – including iridium – would be spread around the planet by the ensuing inferno. They calculated that the sheer volume of pulverised rock stirred into the atmosphere by the asteroid's energy could easily be enough to reduce the amount of sunlight reaching Earth's surface for many years. The darkness would cause massive global environmental change and, crucially, suppress the ability of plants to photosynthesise, thus cutting the global food chain off at its roots. Ecological collapse would swiftly follow. It all seemed to fit.

They published their iridium data and radical hypothesis in the journal *Science* in 1980, and it was met with raised eyebrows from most of the scientific community.[1] It was, put simply, widely disbelieved. After all, if an asteroid impact big enough to cause a planet-wide biological calamity had trashed the Earth in the relatively recent geological past, would one not expect to find more than a one-centimetre-thick layer of sediment enriched in some obscure chemical element like iridium? Would one not expect to find a more obvious trace, like an impact crater, perhaps?

By 1987, seven years after the iridium layer had been discovered, three geologists from the Geological Survey in Denver (USA) discovered an exceptionally rare geological phenomenon named 'shocked quartz' in the K–Pg boundary layer.

Quartz is one of the most abundant minerals in the Earth's crust (sand on the beach is typically made from ground-up pieces of quartz, for example) and it is one of the most widely recognised thanks to its common usage as a semi-precious gemstone. It is also mechanically robust. Shocked quartz is chemically identical to normal quartz but is placed in sharp distinction by its crooked crystal structure, which is warped on the atomic level. Down the eyepieces of a geological microscope, ordinary quartz often appears as a featureless mineral without blemish, but shocked quartz is crosshatched by countless microscopic fractures. Crystalline defilement in a mineral as robust as quartz can only be inflicted by the most severe of external shock pressures. The minerals in the K–Pg layer had suffered great trauma. Indeed, shocked quartz is only forged by two things: underground nuclear bomb detonations, and hypersonic collisions between asteroids and a planetary surface.

The evidence was mounting: the K–Pg mass extinction was caused by an enormous asteroid striking the Earth, and a centimetre-thick layer of sediment that separates the Cretaceous and the Paleogene contained a record of the event. From northern to southern Europe, New Zealand, the American Midwest, and the middle of the Pacific Ocean, the K–Pg boundary layer, complete with its iridium spike and the shocked quartz, was everywhere. The impact must have been truly enormous to distribute fallout across the entire globe.

Like any good hypothesis, this idea explained much of the observational evidence: the global mass extinction; the thin layers of rock enriched with a celestially abundant element; and shocked quartz, which could only have formed during an impact.* And crucially, it made some testable predictions, the most obvious being the hypothesis for the formation of an enormous impact crater. An impact on this scale would leave behind quite a hole, and yet a sixty-six-million-year-old impact crater had never been unearthed.

The hunt for the missing crater was on.

A hidden scar

In the 1950s, exploration geologists searching for untapped oil reserves in Central America had made clear that there lay an enormous circular structure buried beneath the Yucatán Peninsula in the Gulf of Mexico. Two geologists performed a follow-up survey of the area in the 1970s by mapping out slight disturbances in the subterranean rock using advanced ground-penetrating imaging techniques. Their data brought the sweeping circle beneath the Mexican landscape, 180 kilometres in diameter, into sharp focus, but since they had not seen any of the tell-tale signs of an impact from space – such as shocked quartz – they believed the circular structure was the remnants of a long-exhausted volcano.

* Unless the dinosaurs were performing nuclear weapons tests sixty-six million years ago, which seems rather unlikely.

Decades later, however, with the knowledge that there was a missing giant impact crater punched somewhere into Earth's surface, the attention of the geological community turned back towards Yucatán. The two exploration geologists who had mapped out the circle in the 1970s teamed up with another five geologists, and they re-examined samples collected from the area decades before. In the pages of the Yucatán rock record, new and thrilling details came to light that before had gone unnoticed.

In the region surrounding the circular structure, the team discovered a layer of jumbled rock some ninety metres thick that resembled a chaotic layer of baked ash laced with giant boulders. And it was entwined with lobes of quenched melt, too. Tantalisingly, the jumbled layer contained mineral fragments that bore the hallmarks of being traumatised by intense pressure, including shocked quartz. The rock was one of the rarest to grace our planet because it forms only in one specific setting: explosive impacts from space. It takes the name 'suevite' (pronounced 'sway-vite'), and it is forged by the baking and firming up of fallen ash and debris following a giant impact.

Things were rapidly starting to be pieced together: the circular structure; the chaotic suevite; the shocked quartz; the iridium. The team of seven published their findings in the journal *Geology*,[2] and their article was, in stark contrast to Alvarez' iridium paper, immediately met with positivity and enthusiasm. One of the scientists who peer reviewed the piece left a charming comment at the foot of the printed manuscript: '[The authors] propose the long-sought [K–Pg] crater – the smoking gun.' They had indeed found it: they had found the crater that formed on the day the dinosaurs died.

At a dizzying 180 kilometres across and almost thirty kilometres deep (as far down as three Mount Everests), the circular scar remains the second-largest known impact crater on Earth and takes its name from the town at its centre: Chicxulub. It eluded identification for so long because it is simply not an obvious feature of Earth's surface. In the sixty-six million years since the impact crater was formed, the grinding passage of geological time has worn away the circular crater rim and buried the crater beneath a one-kilometre-thick layer of new sediment. Today, it is visible only by using advanced sub-surface imaging techniques.

Judging by the size of the crater and the energy from the blast, the K–Pg asteroid must have been something like twenty kilometres across. That is a stone the size of a city striking the ground at hypersonic velocity. Collisions on that scale release energy otherwise not experienced by a planet, and they far exceed normal geological shocks such as earthquakes and volcanic eruptions. Energy on this scale is impossible to comprehend fully, but a crater that would cover most of the distance between Bristol and Sheffield goes some way towards conveying the magnitude of the event.

Luis Alvarez passed away in 1989 at the age of seventy-seven. If he had lived another two years he would have seen the discovery of the Chicxulub crater. I wish that he had lived to see his hypothesis proven beyond reasonable doubt; I think he would have been thrilled.

Atomic yardsticks

In 1961, the most terrible detonation ever brought about by our species – a hydrogen bomb named Tsar Bomba – was unleashed in the northern fringes of the former Soviet Union. It was 3,000 times more powerful than the Hiroshima bomb that wrought so much destruction and cut short so many lives in 1945. Even with the numbers – a mushroom cloud that reached 70 kilometres into the sky (seven times the height of Mount Everest), windows that were shattered 900 kilometres away from ground-zero, and an earthquake that thrice circumnavigated the planet – it is difficult to comprehend the size of the Tsar Bomba. One way of thinking about it is this: the trigger for a hydrogen bomb like Tsar Bomba is a Hiroshima-style nuclear weapon. Imagine how big an explosion has to be if a 'normal' atomic bomb acts merely as its *trigger*. It is grotesque.

Tsar Bomba is a useful yardstick when considering the energies released by fallen asteroids.

Circles

An impactor around 120 metres in diameter – that is something like the size of the London Eye and just shy of NASA's definition of a 'potentially hazardous object' – would instantaneously release three Tsar Bombas worth of energy upon striking the Earth's surface. A crater some 1.6 kilometres across would be gouged into the crust by such an impactor. Half-a-dozen or so impacts of this scale have afflicted the Earth since we humans began drawing artwork on the walls of caves some 75 thousand years ago.

There are far more tiny asteroids navigating interplanetary space than big ones, and with increasing size, the chance of a collision with Earth rapidly decreases. Consider for a moment asteroids some 300 metres in diameter, around the same size as the Eiffel Tower: they hit the Earth once every 80,000 years on average, meaning that a couple have struck the planet since modern humans evolved. A collision with such an object, which by NASA's standards is formally considered 'hazardous', would scoop a circular crater just shy of six kilometres in diameter into the ground, enough to surround a large town. Forty-six Tsar Bombas worth of energy would be released.

If one of the 900 potentially hazardous asteroids exceeding one kilometre hit the Earth, something like 1,600 Tsar Bombas worth of energy would be liberated in the blast, and a crater some sixteen kilometres across would be gouged into the crust. A crater of this size could comfortably house a large city inside its circular rim. On these scales, even Tsar Bombas are too small a yardstick, which should give us planet-wide pause for thought.

Perhaps the most striking example of an impact crater on Earth is Meteor Crater in the Arizona desert (USA). This bowl-shaped hole stretches some 1.2 kilometres across and would not look out of place on the surface of the Moon. Its circular rim stands proud of the surrounding flat-lying desert before plummeting almost 200 metres down sheer cliffs to the crater floor below. From above, Meteor Crater almost looks as if somebody has dropped a giant ball-bearing into soft plasticine (which is unsurprising, given that is what happened but on a larger scale); from inside, it feels rather claustrophobic, because the horizon is never too far away.

What is more, in the desert surrounding the crater, shards of meteoritic iron litter the landscape, several tonnes of which have been collected over the past several thousand years (beginning with the American Indians, who considered the strange metallic substance sacred). These pieces of meteoritic iron are collectively named Canyon Diablo after a creek that runs by the crater.

The scattered fragments of metal are pieces of the impactor that were ripped away from the main mass as it tore itself to pieces during atmospheric entry. By the time it reached the ground, it was only forty metres across, but being made of iron (which piece for piece is more than twice as heavy as ordinary stone), it left an unmistakable impression on the Arizona desert. Large-scale collection of the curious Canyon Diablo meteorite began in the late nineteenth century, and as fate would have it, one piece wound up in Clair Patterson's chemistry laboratory.

From this piece came Patterson's measurement of primordial lead that allowed him to calculate the accurate age of the Earth for the first time.

Canyon Diablo and Meteor Crater are strange because to find surviving pieces of the offending asteroid near an impact crater is exceptionally rare. Even in the case of the massive K–Pg impactor, bona fide pieces of the culprit asteroid have never been found. Almost always, the sheer energy released by the explosion entirely vaporises the incoming projectile, tearing it to pieces on the atomic level and dispersing it far and wide through the atmosphere. Usually, only chemical traces remain. Shrapnel from the Meteor Crater impactor, though, miraculously survived atmospheric entry, because it rained down far from the site of the violent impact and, since the crater is young, it has not been entirely consumed by the Earth via rusting and degradation in the soil.

Even the craters forged by asteroid strikes are often difficult to find, let alone surviving pieces of the celestial rock. The ever-shifting sands of time quickly wipe them from the surface of the planet and consign them to obscurity: their raised rims are worn to ground level, and their hollow bowls are filled with sediments. Their story is overwritten by new geological tales, but not entirely concealed, and, using the tools of geology, can be gleaned from the ground.

Meteor Crater in Arizona is comparatively small. There are undoubtedly many more craters of a comparable size peppering the Earth's surface that are not so visible, patiently awaiting discovery by inquiring geological minds of the future. Most of the impact craters that we know of show no obvious surface manifestation: no circular rims protrude from the landscape; no bowl-shaped depressions are pressed in the ground; no celestial stones are strewn across the surrounding area. The words written down in the rocks are scribed in a far subtler way, as is the case with the giant impact crater of the Yucatán Peninsula.

Nördlingen

Storks raise their chicks in giant nests atop sloping red-tiled roofs overlooking the medieval town of Nördlingen in Bavaria (Germany), which has been entirely enshrined by a circular fortress wall for over 800 years. At the centre of the town lies St George's Church, whose gorgeous ninety-metre-tall steeple (affectionately named 'Daniel') towers over the tangle of winding streets, and provides a spectacular view across the surrounding farmland. The postcard tranquillity of the Germanic architecture and rolling green fields conceals the tale of disaster written in the rock that lies beneath: the enormous Ries impact crater.

Some fifteen million years ago – less than five minutes ago in our twenty-four-hour geological day – an asteroid smashed into the Bavarian landscape. Judging by the size of the impact crater, the culprit must have been something like one and a half kilometres in diameter, which is twice that of Nördlingen's protective circular wall. Rocks of this size, whilst many times smaller than the giant Chicxulub asteroid, still pack a punch. In a mere instant, 5,500 Tsar Bombas worth of energy was unleashed by the fallen rock, and all hell was raised. Billions of tonnes of stone were subject to temperatures and pressures not normally attained by Earth-bound geological forces, and the sound of the explosion would have rivalled that of heavy traffic as far as 1,000 kilometres away.

Earthquakes tore through the surrounding landscape. Small defects in the rock – perhaps a pre-existing crack or a tiny pebble – acted as points of impedance for the shockwaves, causing them to ricochet around them as water does around the prow of a ship. Deflected outwards and onwards by the miniature defects, the pressure fronts fractured the rock to produce ice-cream-cone-shaped breakages, with the tip of the cone emanating from the defect. These conical fractures, called 'shatter cones', offer smoking-gun evidence for an impact event, because they are only created by the most extreme and instantaneous pressures. The only other way to make a shatter cone is by detonating an underground nuclear bomb.

Shatter cones do more than metaphorically point towards the impact crater: they literally point towards it. The tip of a shatter cone always points backwards towards the direction from which the seismic shockwave came, like little fist-sized arrows in the rock.* By mapping the directions in which the shatter cones across the Bavarian landscape point, geologists have pinpointed the epicentre of the impact to a few kilometres northeast of Nördlingen. There, true mayhem was unleashed and, as with the rocks of the Yucatán Peninsula, the rocks surrounding Nördlingen experienced the most devastating misfortune.

When the first shrapnel made landfall, it scraped and rolled away from the site of the impact, scratching long scores into the ground like nails along a chalkboard. Anything in the way of the falling boulders was obliterated. Subsequent debris buried the scratches within seconds, and on exposed surfaces where the debris has today been swept clear, the scratches are still visible. They are some of the rarest and most bizarre geological structures I have ever seen: radiating outwards from ground-zero like the spokes of a bicycle, they were etched into the floor in a split second.

A scorching column of ash, riddled with lobes of molten stone, would have towered over the destruction and, in the hours following the explosions, settled as a grey-beige coating of

---

* I am the proud owner of a beautiful shatter cone that I found as an undergraduate student during fieldwork in the Ries impact crater. It has been on my bedside table for six years now.

suevite.* Many of the historic buildings in Nördlingen, including St George's Church (and the Daniel tower), are made from
blocks of the stuff. Nördlingen is not just built inside an impact
crater: it is in part built *from* an impact crater.

Unbeknown to the architects and craftsmen of the town,
their buildings are from one of the rarest types of rock on the
planet. They had no way of knowing at the time that they were
using such bizarre rock and no way of knowing what wonders
lay inside the stone.

When the shockwave from the impact rippled through the
ground, atoms in the rocks were squeezed together in awkward
and unusual ways, and, in a brief moment, locked together to
form new and unusual forms of matter. Newfangled and exotic
minerals, like shocked quartz, were crafted. Amongst them was
diamond, too. Countless in number, the diamonds lace the suevite
building stone, perceptible with the aid of powerful microscopes.

Nördlingen really is a town like no other.

Tektites

Fallout from the Yucatán and Bavarian catastrophes was by no
means confined to the landscape immediately surrounding their
impact craters. Some of the molten ejecta was splashed into the
upper stratosphere, reaching some forty kilometres in altitude
(around four times higher than a cruising passenger aeroplane),
and made landfall far away from the site of the impacts. As they
sailed along ballistic trajectories through the atmosphere, the
prune- to salt-grain-sized droplets quenched to form globs and
strings of glass. They froze into many weird and wonderful

---

* I also have a piece of Nördlingen suevite on my bedside table, next to my
shatter cone.

shapes: some are marble-like orbs, while others take the form of squashed spheres; some are finger-shaped rods that taper at either end, while some are stretched dumbbells. Others look like gnarled plasticine. These strange geological entities are only crafted by impacts from space, and we call them 'tektites'.

Within one year of the K–Pg iridium spike's discovery, a pair of geologists from the Geological Institute (Netherlands) discovered microscopic tektites lacing the K–Pg boundary layer in Spain. The tiny droplets of quenched melt, which at less than one millimetre across would comfortably sit on the head of a pin, had travelled thousands of kilometres through the atmosphere before making landfall. So too have similarly tiny tektites been discovered stippling the K–Pg boundary layer in North Dakota, some 3,000 kilometres from the site of the impact, and in 2019, a team of palaeontologists discovered microscopic tektites embedded in the gills of fossilised freshwater fish.[3] Like vicious artillery, the microscopic tektites rained from the skies in the hour following the impact to the south, peppering the animals of North Dakota before they were consumed by the fires and earthquakes.

Most tektites are pitch black with a glassy lustre, but in the case of the European tektites from the Ries impact, the chemistry of the molten Bavarian rock is such that they quenched to form translucent bottle-green glass. Their unique colour and lustre distinguishes them from all other tektites and bestows upon them a special name: 'moldavites'. They are prized amongst jewellers and mineral collectors. Moldavites are found to the east of the Ries impact crater – usually in the Czech Republic – but curiously, they are not found in the west, north, or south. The implication of this simple observation is that the asteroid which smashed into Bavaria fifteen million years ago must have entered the Earth's atmosphere on a shallow angle from the west, throwing most of the fallout downrange towards the east.

The vast distances covered by the tektites and the stunning atmospheric height that they reached is a testament to the raw power of the explosion. Moldavites have been found further than 300 kilometres from the Ries impact crater, which is further than the distance from Glasgow to Leeds. But horrifyingly, stray asteroids are not the only things capable of forging tektite-like substances.

On 16 July 1945 at twenty-nine minutes and forty-five seconds past five in the morning (local time) in the New Mexico desert, the United States successfully demonstrated technology with enough power to spell an end to human civilisation. The first nuclear weapon of mass destruction, code-named 'Trinity', was detonated. It marked one of the most significant turning points in the 200-thousand-year history of our species and was the first moment that we became an existential threat to ourselves.

The Trinity nuclear bomb was powered by a nuclear chain reaction that released the equivalent of 21,000 tonnes of TNT in less than one second (which, by the nuclear-bomb standards of today, is rather puny). The choice of fuel was a mixture of plutonium-239[4] and uranium-235. Upon exploding, the core of the bomb, which weighed a measly six kilograms, created a mushroom cloud hotter than the surface of the Sun that bloomed twenty kilometres into the sky.

Enrico Fermi, a scientist who worked on the bomb's development as part of the Manhattan Project, described the destruction at ground-zero: 'a depressed area 400 yards in radius glazed with a green, glass-like substance, where the sand had melted and solidified again'.

In the vicinity of the detonation, the top layer of desert sand was instantly molten and fused to form an eerie green glass. It has since become known as 'trinitite'. It formed by similar means to the tektites — instant and complete melting of rock, followed by rapid quenching – but instead of being created by a stray stone from space, it was created by us. We did it.

We humans have been pressing our footprints in soft sand for more than 200,000 years, etching patterns on lumps of rocks for at least 75,000 years, and creating beautiful murals on the stony walls of caves for at least 40,000 years. Trinitite is just one of many examples of the power we humans have to shape the Earth and leave our mark in the rock record. At the same time, it is a reminder that right now, today, we also hold the power of falling asteroids.

Christmas Eve

Not all stray celestial stones, reassuringly, have the power of the K–Pg asteroid or Canyon Diablo, and most witnessed meteorite falls cause a mild commotion rather than a planetary-scale extinction event. Of the 60,000 or so known meteorites, fewer than 1,300 have been seen falling. Most falls were tiny and hit the ground with nothing more than a dull thud, but occasionally, as was the case with meteorite falls like Allende of 1969 and Ensisheim of 1492, one lands with a little more flare. Another meteorite that caused a commotion did so over the English Midlands in 1965.

'Twas the night before Christmas when a bright fireball was seen streaking across the skies above Leicestershire through gaps in patchy clouds.* It was an ordinary chondrite (like most meteorites), around the size of a turkey, that had been swept up by the Earth after an interplanetary voyage from the Asteroid Belt. The celestial stone broke into pieces as it plummeted through the sky, and its fiery flight culminated in the small village of Barwell.

---

* Imagine what the children of Leicestershire must have felt as they saw Father Christmas falling from the sky in a ball of flames.

Residents who were outside at the time reported a loud bang accompanied by a flash in the sky, followed swiftly by a swoosh and the sound of heavy objects making landfall in the streets. It was raining pieces of the Barwell meteorite: stones smashed through living-room windows, damaged roof tiles on houses, and punched potholes in the road. One Barwell resident left his house to investigate the commotion and found his brand-new Vauxhall Viva had been struck by a falling stone. The insurance company refused to pay out, describing it as an 'act of God'. Perhaps putting it down to an 'act of gravity' may have been more accurate, but would doubtless have been little comfort to the distraught owner. There are rumours that he asked a local priest what God had to say about his damaged car, requesting that the Church foot the bill. His plea, apparently, was denied.

Residents quickly collected pieces of the stone, most of which were sent to the Natural History Museum in London for curation and scientific research. The Barwell meteorite fall remains the biggest witnessed fall in Great Britain and its arrival is commemorated by a green plaque in the village, set in a large (non-meteoritic) boulder.

Meteorite falls over residential areas are incredibly rare, but then again we humans do occupy only a tiny slither of the Earth's surface. Most meteorites fall into the ocean and of those that do fall on land, most do so without being spotted. Damage to property has not often occurred and only a few collisions with a human have ever been documented.

One unlucky victim of a falling meteorite was Ann Hodges of Alabama (USA). She was having an afternoon nap on her living-room sofa in 1954 when a meteorite crashed through the roof of her house and hit her radio before bouncing into the side of her body, leaving a bruise the size of a rugby ball. The meteorite, classified as an ordinary ('H') chondrite, was officially named

Sylacauga after the nearest big city, but it is more commonly referred to as the Hodges meteorite.

There has only been one recorded death caused by a falling meteorite.[5] In Iraq on 22 August 1888, a meteorite entered the Earth's atmosphere. It exploded in mid-air and broke into pieces. Incredibly, two of the lumps hit two unsuspecting victims, paralysing one and killing the other. Both men won a terrible cosmic lottery.

Thankfully, of the 40,000 tonnes of rock that falls to the Earth's surface each year, most is in the form of minute fragments. Even the Barwell meteorite, which was an uncommonly large fall, only gouged small pits into the surface of the roads and pavements. But just occasionally, monstrous pieces of asteroid shrapnel happen upon the Earth, and do untold damage.

Rush-hour fall

A not-so-gentle reminder of the raw power of even a modest-sized asteroid was laid bare for the world to see in 2013.

It was twenty past nine in the morning on Friday, 15 February 2013. The crystal-clear winter skies above the small Russian town of Chelyabinsk had been steadily brightening, and the snowy landscape was flooded with morning sunlight. The hum of car engines and the slamming of car doors sounded in the freezing air. Townsfolk were going about their morning routine – driving to work, dropping children off at school, taking out the rubbish – when a small ball of light appeared. It grew brighter, and brighter, and brighter, and left a thick white tail in its wake as it traversed the sky. The shards of light made the winter's day seem soft by comparison.

Brighter it grew, and larger, until it was no longer possible to look at it without squinting. Then it exploded. For a few seconds, the flare burned brighter than midday sunshine, and it cast light into the skies in every direction for 100 kilometres. The snow-covered landscape was momentarily illuminated with the light of thirty Suns, and sharp shadows were fleetingly cast by trees, buildings, and people upon the snowy ground.

Then came the sound. *Boom.* Echoes of the explosion reverberated across the wintery Russian landscape and the shockwave shook the earth. The terrible thundering noise seemed to last for a lifetime. *Boom.* Windows were shaken from their frames and reduced to glassy shards; a factory roof three times the area of a tennis court collapsed; people were shaken off their feet and sent tumbling to the ground. *Boom.* As the noise faded into a rolling rumble, the only sounds left were choruses of car alarms and the dazed scrambling of the bewildered townsfolk. Some £25 million worth of damage had been inflicted upon buildings in under five seconds. The whole thing was over in around half a minute.

In a sentiment reminiscent of the Argentinian legends of the Campo del Cielo Field of the Sky meteorite that had fallen some four thousand years before, residents of Chelyabinsk likened the event to the Sun falling from the heavens. In reality, another enormous stone had just touched down on planet Earth. The Chelyabinsk meteorite, the largest natural object to fall from space in over a century, instantly made international news.

The mayhem had been caused by an ordinary chondrite the size of a seven-storey building travelling at twenty kilometres per second. When the twenty-metre lump hit the Earth's upper atmosphere it immediately began decelerating. Twelve thousand tonnes of stone, the same mass as 3,000 adult elephants, slashed through the sky, and when it reached a height of thirty kilometres above the ground, it tore itself to pieces. An almighty explosion ensued. The amount of energy released in the detonation is startling: thirty times more than was released by the atomic bomb detonated over the city of Hiroshima in 1945. The shockwave from the explosion injured almost one thousand people in the town of Chelyabinsk before circumnavigating the entire globe.

Fragments of the stone, each coated in a liquorice-black fusion crust, peppered the frosted landscape. Most were small, no bigger than the fist-sized pieces of the Barwell meteorite, but some were truly enormous.

As it fell, the main mass was, by chance, captured on video by security cameras and a fleet of dash cams from different angles. The multiple perspectives from which the stone's flight was captured allowed its path to be triangulated and accurately calculated, and so its trajectory could be projected forwards. This gave scientists and meteorite collectors alike a good idea of exactly where the giant stone had come to rest.

The landing site was narrowed down to the area around Lake Chebarkul some seventy kilometres west of Chelyabinsk. Excited search parties quickly arrived on the scene expecting to find a giant black rock resting awkwardly atop the snow-covered landscape, but no such rock was found. Traces of the meteorite soon became apparent, however. An enormous hole, eight metres across, had been punched into the ice atop the frozen lake: the meteorite had crashed through the ice and lay submerged at the bottom of the dark, icy waters.

Months later, and after much thawing of the lake's icy surface, a search team pinpointed the meteorite's resting place with the aid of sonar and expert divers. Then came the recovery mission. Exhuming an enormous stone from the bottom of a silty lake is no easy feat, and the local authorities employed the help of many divers and a powerful mechanical winch. Eight months later, the stone was raised from the watery depths in the presence of a lively crowd of locals and journalists, and the event was broadcast live on television. In a rather undignified manner, the giant stone, boasting a blackened crust of fused rock, was hauled onto the lake's shore.

By meteorite standards, it was truly enormous. The stone tipped the scales — literally — when it weighed in at over half a ton, and broke into three large pieces. Photographs from the day paint a picture of a frenzied scene of excitement and utter chaos. A few pieces more were recovered from the floor of the lake and took the total amount of recovered meteorite to something approaching one tonne, making it the sixth-largest fall on record. Thanks to the proliferation of dash cams in Russia and the fortuitous location and timing of its fall, the Chelyabinsk meteorite is perhaps the best documented of all meteorite falls. It captured the imagination of the world.

When the K–Pg asteroid vaporised upon striking the Earth's surface, the chemical medley from which it was made was driven high into the atmosphere and distributed across the planet. This is the origin of the enriched iridium in the K–Pg boundary layer. But the chemical traces of the impact are just the beginning, because Nature also scribed the K–Pg tale of catastrophe in the language of isotopes — specifically, isotopes of element number twenty-four, chromium.

Similar to the way that the isotopes of oxygen can be used as powerful tools to organise meteorites into groups, so too can the isotopes of chromium. All chromium atoms have twenty-four protons in their atomic nuclei (by definition), but their varying numbers of neutrons give rise to a total of four different isotopes, which vary in proportion systematically and predictably. Meteorites, it will come as no surprise to hear, break this rule, because they originate from fundamentally different planetary bodies that inherited unique blends of isotopes from their dusty nebula building blocks. Most groups of meteorite have a distinctive blend of chromium isotopes.

It turns out that the blend of chromium isotopes in the K–Pg boundary layer is distinctly non-Earthly – a testament to the presence of vaporised asteroid that was mixed into that particular layer of rock. By measuring the precise proportions of these isotopes within the boundary layer and comparing them to the wide array of meteorites we have at our disposal, a team of cosmochemists at the Laboratoire de Géochimie et Cosmochimie in Paris deciphered the isotopic character of the impacting asteroid.

They discovered that the asteroid that caused the mass extinction at the K–Pg boundary was a water-bearing type of meteorite that is brimming with organic chemistry to the point of stench. The asteroid that ended the reign of the dinosaurs and extinguished so many other forms of life was a giant carbonaceous 'CM' chondrite, similar to Murchison.

In a clever twist of irony, the same celestial stones that may have delivered the sparks of life to our planet also caused one of the greatest of mass extinctions in geological history. Asteroids and meteorites are not only the likely givers of life; they are the takers of life, too. But from destruction comes chaos, and in chaos lies potential. The gaping hole left behind by the mass extinction represented a golden opportunity for evolution to steer new forms of life into being.

With the demise of the dinosaurs, a chance arose for a group of small burrowing animals to leave their holes in the ground. These tiny warm-blooded creatures, which are unique in the animal kingdom in that they give birth to live young, rear their infants on milk, and have fur, quickly shrugged off the apocalypse and flourished. They are, of course, the mammals, and their rapid diversification in a post-extinction world would, after sixty-six million years, lead to the wonderful assortment of mammalian characters that inhabit the Earth today – from bats to whales, from cats to humans, and everything in between. All of us warm-blooded, milk-drinking, fur-clad beings descend from ancestors who took the opportunity to fill an ecological niche blown into existence by an asteroid smashing into the Yucatán Peninsula.

Without the carbonaceous meteorites raining star-tar upon the Earth, the sparks of life may never have caught; without the carbonaceous asteroid sending Earth's ecosystem into ruin sixty-six million years ago, life would have taken a path that would never have led to the proliferation of the mammals, and eventually to us. Meteorite impacts are simply one of Nature's ways of pruning the tree of life and allowing new shoots to grow. We owe our existence to them in more ways than one.

Upwards and onwards

The destruction of rock is as much a part of geology as the creation of rock. Solid stone – the solid ground beneath our feet, which feels as though it has existed for ever and will outlast eternity – will eventually crumble. Nature operates with absolute indifference. Nothing, in the end, is permanent. Entities such as stone formed by the coming together of atoms are ever-shifting, fleeting features of the cosmos.

Geology is a force that makes use of the material it already has, and every time a rock is unmade a new one is forged in its place. Destruction is the prerequisite for creation; both are intimately linked in an ever-turning cycle. Yin and yang; chaos and order. Whilst they appear opposites, they are ultimately part of a whole, each complementing the other as the creaking wheel of time slowly records geological history into solid stone.

Towering cliffs of three-billion-year-old rock are slowly reduced by the sea to sand on the beaches of today; those grains of sand will make up the sandstone layers of tomorrow. Trees transform soil, water, and air, into wood, roots, and leaves, before decaying into soil once more. Our living bodies are much the same: we will one day be unmade and returned to the Earth from whence we came, and our parts will be made into entities anew. From rock we came and to rock we will return.

Apocalyptic events have continually punctuated the otherwise steady passage of geological time on our home planet. The rocks of Earth reveal stories of utter ruin: events that shook the Earth but left in their wake a rich tapestry of exotic minerals, strange chemistries, and unique rock types. If they are large enough, stray celestial stones write the most calamitous tales, but from them, new geological forms of beauty and intrigue emerge, and some have the power to shape the course of life itself.

Having all of the human beings in the Universe – and all known life, for that matter – confined to one world is (at best) precarious and (at worst) catastrophic. As the saying goes, it is unwise to place all of one's eggs in a single basket. All it would take is one rogue asteroid of sufficient size and our civilisation would be rocked to its core. Millions of lives could be lost if the impactor made landfall within a few hundred kilometres of a major city, and billions of pounds worth of infrastructure destroyed. Atmospheric cooling, triggered by the dust ejected high into the atmosphere, could cause food scarcity across the globe for a decade or more.

On top of the direct loss of life, entire regions could be politically and economically destabilised. Large-scale human displacement as a consequence of such an event would be a humanitarian crisis that dwarfs anything we have yet experienced as a global civilisation.

Asteroid deflection is one way we could avoid such a grim scenario and is probably a strategy we will deploy in the far future when we discover an asteroid with an Earth-crossing orbit. But deflection of asteroids is itself fraught with dangers and is a course of action we must not take lightly. A mistake could mean that an asteroid was deflected in such a way that it became more likely to hit the Earth. It is also not beyond the realm of possibility that such technological capability could be grossly misused by malevolent forces: if we had the power to deflect asteroids away from the Earth, we would also have the power to deflect asteroids towards it. History has taught us that individuals sometimes come to power who would jump at the chance of wreaking such destruction. We must think carefully before we grant ourselves such technological prowess.

We are a long way away from shifting the orbits of asteroids and directing them away from the Earth. But we are not a long way away from another solution to the problem: interplanetary settlement. It may well be the case that it is easier and wiser to settle on the surface of other worlds than it is to deflect asteroids.

Meteorites offer a persistent reminder that the planet beneath our feet is not the only surface made of stone in the Solar System. There are countless other worlds, other places out there, which we could choose to stand on if we so chose. There are many we could feasibly settle on – and eventually flourish on – given a few more decades of technological innovation.

We have already made steps towards this end by setting feet on the surface of the Moon. And, after almost two decades of continued habitation onboard the International Space Station, we have also shown that humans are capable of living (and thriving) in micro-gravity environments. The prerequisites are there; now all we need is the will.

Interplanetary human space flight, and eventual habitation of other rocky worlds, lies in our future. If we are to avoid human extinction via a rogue asteroid, it absolutely must. Upwards and onwards we must sail.

The stones that fall from the sky were there at our story's beginning; they have shaped our story since; and they will continue to do so as it stretches ever on, filling the blank pages with tales unwritten.

Epilogue: The Story Goes On

Following Earth's story backwards, we have traced it from the depths beneath our feet and upwards into the sky. Meteorites, the rocky pages that blew across the void of interplanetary space from other worlds, tell us the first chapter. Taking us further backwards in time than any Earthly rock, these celestial stones relay the earliest history of our Solar System and the worlds that grace it.

As beings of the Earth, this is part of our own story, too. We humans have something like five thousand years of written history, which was preceded by tens of thousands of years of pictographic history, which was preceded by billions of years of evolutionary and geological history. There is an unbroken chain of events that ties each of our 'here's and each of our 'now's to the cosmic assembly of our Solar System 'then', some 4.6 billion years ago. We all share a common heritage.

Small pockets of this interstellar cloud – through stellar, physical, chemical, then geological, and then biological, evolution – eventually came to gaze outwards into the cosmos and contemplate its own story: us. We, small pieces of a nebula come to life, have plumbed the abyss of deep time and discovered the natural history of our Solar System and of ourselves. We have discovered a story written across light years down the barrel of a microscope, and come face to face with the grandest of timescales.

The story is still unfolding. Blank slabs of stone pave the way into a future unknown. New stories will be etched into new rocks, and new histories will be written.

But the story from here onwards, for both us humans and planet Earth, will be different from the chapters that came before. The words that are written into the next few pages of Earth's story will, for the first time, be in part consciously written by one particular species: *Homo sapiens*. Us. We can create new geological materials like trinitite; we can blanket the seabed in exotic materials like plastics that will become part of new rocks; we are changing the climate on a planetary scale, and at a rate that far exceeds the natural order of things.

We live in a world we are told has never been more divided, and on a planet we are told is in ecological and environmental turmoil. It is little wonder that things seem never to have been more hopeless. We hold ourselves to high standards, and rightly so: we are capable of the most incredible feats of ingenuity and humanity, yet somehow we often miss the mark.

Yet still, despite it all, there is much cause for optimism and hope.

The *Apollo* space programme of the 1960s is a pertinent reminder that if there is a problem and the will exists to solve it, there is probably a way to think and invent our way out of it. Our species went from launching comparatively primitive bleeping satellites in 1957 to round trips to the Moon carrying astronauts in 1969. That is little over a decade! And since then, we have soft landed robotic space probes on the solid surfaces of eight celestial bodies: two planets (Venus and Mars), our Moon, Saturn's giant moon Titan, comet 67P/Churyumov–Gerasimenko, and three asteroids (Eros, Itokawa, and Ryugu). We have placed spacecraft into orbit around six planets; we have ollected and brought back the solar wind from the Sun and dust from the tail of a comet; we have had a permanent human presence in space for almost two decades onboard the International Space Station; we have performed a reconnaissance of the entire Solar System with spacecraft such as the *Pioneer* twins and the *Voyager* twins; we have discovered the Universe with new eyes through advanced telescopes like the Hubble Space Telescope and the Arecibo radio telescope. It is not unreasonable to expect that by the time you are reading this book, we will have successfully touched down on a ninth celestial body: Bennu, an asteroid near Earth, laced with star-tar and water-rich minerals.

Science continues to give our species an excuse to be the best versions of ourselves. We can do anything that we collectively put our minds to.

We are capable of adding our own forms of beauty to the Universe by understanding the story of how we came to be, through using our powers of thought and experimentation. Comprehending how the stars shine; understanding how the trees grow from thin air out of the ground; viewing the world as an assembly of ninety-two or so chemical elements; reading the geological story, which saw its origins high in the celestial sphere, of how we came to be. We did all of that with our minds and our hands.

One of the most exciting things about science is how long (or rather, *not* long) we have been practising it. The written word has been around for 5,000 years; we have only been practising the scientific method for around one-tenth of that time; and we have only been aware that we have the power to change the Earth on a planetary scale for around one-tenth of even that time. Science is a new invention, and so we should not be too surprised if, on occasion, we slip up and make mistakes. Even grave ones. This is not to excuse the irresponsible use of powerful technology. Rather, it places the burden of self-reflection and responsibility squarely on our shoulders and forces us directly to confront how good things could be should we will them into existence. If that is not motivation to act responsibly and to strive to solve the problems that face us, then I do not know what is.

Despite the mistakes we have made during our brief tenure on this planet as a scientifically literate technological species, we humans still contribute a great value to the Universe simply by existing. We are pieces of the nebula – pieces of the Solar System and the Universe – looking at ourselves, outwards through telescopes, inwards down microscopes, and backwards into the canyon of deep time at the path down which we have walked. We are, at least on the Earth, the only forms of matter capable of contemplating their own existence and understanding their own story. The eyes of every individual are a way for the light of the world to enter a mind capable of knowing itself. We are the only way we know of that the Universe can appreciate its own beauty.

Blessed with the knowledge of where we came from and the fruits borne by scientific enterprise, we have both the imperative and the means to survive into the far future, a future in which we will right the mistakes we have made, and one in which we will continue to reach new heights. There is no reason that this future is not only possible, but likely. We must not sell our species short. We owe our future existence to ourselves and the Universe, because we are a part of the Universe.

From gas to dust; from dust to worlds; from worlds to minds. There was a time when we were not; there is a time, *now*, when we are; and there will be a time when we will not be. Earth's story will go on regardless of whether we are a part of it or not. Our past is written in rock – Earthly and meteoritic alike – but our future is not.

Now: over to you.

Appendix: Meteor Showers

It is impossible to predict where a meteorite might fall and to see it with your own eyes, but other stones from outer space that never make landfall can be viewed in the skies as regularly as clockwork year after year.

Some of the ice-rich planetesimals that survived the era of planet formation some 4.6 billion years ago persist to this day. Many are in elongated – as opposed to roughly circular – orbits: they sweep from the cold outer areas of the Solar System into the boiling hot inner regions, quickly swing around the Sun, before being catapulted back into the cold. They repeat their journey, retracing their path, orbit after orbit after orbit. We call these celestial objects 'comets'.

When close to the heat of the Sun, their surface ices (e.g. water, ammonia, methane, carbon dioxide) and other volatiles vaporise. Their exteriors roil as angry jets of vapour hiss forth. Pieces of rock, most no larger than a grain of sand, are entrained in the fountains of vapour and shed into the vacuum of space. These minute pieces of rock are 'meteoroids'. Many meteoroids are left in a comet's wake.

With each orbit, more rock is blasted from a comet's surface, and the trail of meteoroids grows thicker. Spreading out along the orbit of the comet, this rocky debris forms a thin ring around the Sun called a 'dust trail'. Each comet produces its own dust trail. They are responsible for one of the most celebrated and captivating of all astronomical events: 'meteor showers'.

During its annual orbit around the Sun, the Earth sometimes ploughs through a comet's dust trail, and when the meteoroids enter the Earth's atmosphere, they burn up because of their breakneck speed. They are travelling so fast that they glow incandescent. We call these streaking lights 'meteors', but they are more commonly known as 'shooting stars'. They are visible even against the orange glow of our heavily polluted modern night skies. Meteors shine incredibly bright.

Meteors streak across the sky every night of the year,* but their numbers increase, sometimes dramatically, during times when the Earth is sweeping through a cometary dust trail. These ephemeral spikes in meteor frequency are called 'showers'. Each particular meteor shower happens at the same time each year because the Earth sweeps through a particular cometary dust trail at the same point in its orbit (i.e. the same calendar date).

From our vantage point down here on the ground, meteor showers appear to radiate outwards from a single point in the sky, when in reality they travel along parallel trajectories. It works like this: imagine you are stood between the tracks of a disused railway line that stretches forth in a straight line over the horizon. The two railway tracks, which are exactly parallel, appear to converge on a single point as they approach the horizon. Much the same happens with meteor showers but in reverse: they appear to radiate outwards from a single point purely because of our perspective point. Each shower takes its name from the constellation in the sky from which it appears to be radiating (e.g. the Perseids radiate from the constellation of Perseus).

---

* And during the day, for that matter, although they are invisible in bright daylight.

There are more than one hundred annual meteor showers which range in intensity from around half a dozen meteors per hour to one hundred per hour or more, although individual showers range in intensity from year to year. On occasion, the Earth sweeps through an unusually dust-rich region of a comet trail, and a meteor shower can reach epic intensities. We call particularly spectacular showers 'meteor storms'.

No specialist equipment is required to see them. Telescopes are of little use because meteors trace out long paths across the sky rather than a single fixed point. Accessible to all, the only things one needs to enjoy a celestial light show is a clear view of a cloudless sky, patience, and a warm coat or blanket (and flask of tea). Fortunately for early risers, meteor showers reach their highest intensities in the hours between midnight and dawn, when we are on the side of the Earth facing 'forward' in its orbit: in this time window, meteoroids slam into the atmosphere head on, and so appear much brighter.

Below is a compilation of the most spectacular annual meteor showers for your enjoyment, the dates over which they occur, their intensity, and the comet that is responsible. Happy watching.

Shower	Constellation	Dates*	Peak*	Intensity	Comet
Quadrantids	Boötes	22 Dec.–17 Jan.	4 Jan.	Very high	(196256) 2003 EH₁**
Lyrids	Lyra	14–30 April	22 April	Low	C/1861 G1 (Thatcher)
Eta Aquarids	Aquarius	17 April–24 May	6 May	High	Halley's Comet
Southern Delta Aquarids	Aquarius	21 July–23 Aug.	29 July	Low	96P/ Machholz
Perseids	Perseus	17 July–1 Sept.	12 Aug.	Very high	Comet Swift–Tuttle
Orionids	Orion	23 Sept.–27 Nov.	22 Oct.	Low	Halley's Comet
Leonids	Leo	2–30 Nov.	18 Nov.	Low	55P/ Tempel–Tuttle
Geminids	Gemini	1–22 Dec.	14 Dec.	Very high	3200 Phaethon
Ursids	Ursa Minor	19–24 Dec.	21 Dec.	Low	8P/Tuttle

* These dates may move around slightly from year to year due to leap years, etc.

** (196256) 2003 EH₁ is technically an asteroid, highlighting the blurry boundary between what we classify as 'comets' and what we classify as 'asteroids'.

Acknowledgements

All works of science are collaborative, and behind every name attributed to a discovery there are a myriad of contributors. I discovered this first hand while completing my MEarthSci and PhD; writing this book has been much the same. I have many people to thank.

First and foremost, thank you to my Mum, who raised my sister and me single handed. She made sure we always had a peaceful place at the kitchen table where we could do our homework and a bookshelf stacked with encyclopaedias, has always fostered our interest in the natural world, and she continues to give her full support to everything we do in life.

Thank you, Lucy Kissick, my partner in leaf, best friend, and dedicated proofreader, who inspired me to put pen to paper. Without her love and support this book would never have been written; she has been there from the first word to the last. Every day I feel lucky that we are alive in the same thin slice of deep time. 'There are roads left in both of our shoes.'

Thank you, Lucy Manifold, for being there during the best and the worst of times, and for always having a plan; Mr Currie, for rekindling my love of geology; 'Doctor Anne' Hoyle, for sound advice; Mrs Thompson, for showing me the importance of literature; Tim Elliott, for unwavering support; Steve Noble, for raising my spirits and keeping me plied with Yorkshire tea; Chris Coath, for being wonderful to work with; Sara Russell, for infectious enthusiasm; Jamie Gilmour, for teaching me that there is more to getting a degree than getting a degree; Katie Joy, for showing me the joys of achondrites; Mike Zolensky, for introducing me to meteoritics; Loan Le, for cheering me on across the Atlantic; the School of Earth Sciences (University of Bristol), for being the best place to have completed my PhD; the British Geological Survey, for being a wonderful place to be a research scientist; Danny Stubbs, for always being a part of the excitement; Dan Bevan and Madds Thornton, for escapism in the Tranquillity Base; Sophie Williams, for reminding me of home; Aivija 'Isotope' Grundmane, for helping me with my organic chemistry; Diksha Bista, for being a delightful labmate; Mrs Brougham and Mrs Hirst, for helping me find my place during secondary school; Mr Guest, for introducing me to the geological delights of North Wales; Mr Ryder, for opening up the world of music to me; Mr Ellis, for the honest advice which has served me well for over a decade; Mr Gilroy, Mr Naylor, and Ms Owen, for nurturing my love of science.

And when it came to this book: thank you, Georgina Laycock, for believing in me and making my first publishing experience a joy; Candida Brazil, for her careful edits; Howard Davies, for his incredible care and eye for detail; and Caroline Westmore for giving *Meteorite* an interstellar diamond polish; Northbank Talent Management, for helping me share my science with the world, and especially Martin Redfern, for making *Meteorite* a reality; Helen Thomas, for encouraging me to spin a yarn or two; Sue Rider, for suggesting I write it in the first place; and my beloved cats Missy, Peaches, Priestley, and Shelly, for keeping me company while I wrote.

And finally, thank *you*, dear reader. Inquisitiveness and wonder will always serve you well. Stay curious.

Notes

Chapter 1: Stones from the Sky

1. Comelli *et al.* (2016), *Meteoritics & Planetary Science*, vol. 51, pp. 1301–9.
2. Topham (1797), *Gentleman's Magazine*, vol. 67, pp. 549–51.
3. Howard (1802), *Philosophical Transactions of the Royal Society of London*, vol. 92, pp. 168–212.
4. Herschel (1802), *Philosophical Transactions of the Royal Society of London*, vol. 92, pp. 213–32. It is a happy coincidence that Howard's cosmochemical analyses and Herschel's coining of the word 'asteroid' appeared in the same volume of the same journal.

Chapter 2: Earthfall

1. Meteoritical Bulletin Database, as of early 2020. The Meteoritical Society, an international academic society founded in 1933 to promote and support research into meteorites, maintains the Meteoritical Bulletin Database. The online database compiles and catalogues all known meteorites and is updated regularly. It is freely accessible at www.lpi.usra.edu/meteor. I use it all the time.
2. We bought the apple-sized piece of meteorite that I studied for almost four years as part of my PhD research from ... wait for it ... eBay (specifically, a reputable meteorite dealer by the name of 'Mr Meteorite'). I imagine my purchase raised a few eyebrows in the university finance office when the research expense form came across their desk.
3. Clayton *et al.* (1973), *Science*, vol. 182, pp. 485–8.

4. Antarctic meteorites are named after the place on the ice sheet where they are found, plus a (usually) five-digit identification number. The first two digits represent the field season in which it was found, and the last three digits represent the order in which the stones were catalogued at the Johnson Space Center. Allan Hills 81005 was discovered in Allan Hills, during the 1981 field season, and was the 5th stone to be catalogued in Houston.

5. This can be illustrated using some relatively simple maths.

    If we doubled the diameter of an asteroid, its volume would increase by a factor of $2^3 = 8$, while its surface area would only increase by a factor of $2^2 = 4$. If we tripled the diameter, its volume would increase by a factor of $3^3 = 27$, while its surface area would only increase by a factor of $3^2 = 9$. Volume increases rapidly compared to surface area.

    Here is an example. Suppose we compare two spherical asteroids, one that is 3 km across and one that is 25 km across. The equations we need are: surface area of a sphere $= 4 \times \pi \times \text{radius}^2$, and volume of a sphere $= (4 \div 3) \times \pi \times \text{radius}^3$. (Remember, the diameter of a sphere is twice the radius.)

    The surface area of the smaller asteroid is ~ 28 km^2, its volume is ~ 14 km^3, and so its surface area to volume ratio is ~ 2 ($28 \div 14$). The surface area of the larger asteroid is $\sim 2{,}000$ km^2, its volume is $\sim 8{,}000$ km^3, and so its surface area to volume ratio is $\sim \frac{1}{4}$ ($2{,}000 \div 8{,}000 = \frac{1}{4}$). Larger asteroids have a smaller surface area to volume ratio.

    The exact same calculation could be used to show that a cup of tea in a large mug would stay hotter for longer than a cup of tea in a small mug.

Chapter 4: Spheres of Metal and Molten Stone

1. The Widmanstätten pattern is named after Count Alois von Beckh Widmanstätten, an Austrian scientist who discovered this unique mineral structure in 1808. It had in fact been discovered by English scientist William Thomson four years earlier. Thomson published his findings (along with a clear sketch) in a non-English journal but died two years later. This, unfortunately, meant that his discovery did not become known far and wide.

2. These colours, exhibited by most rocks when viewed in thin section, are called 'interference colours'. When a beam of white light is shone through a thin section, certain wavelengths of light are stripped out of the spectrum by optical interference as they interact with the molecular structure of the minerals making up the rock. This changes the colour of the light transmitted out of the other side of the thin section and into the human eye (or a camera lens). The colours are not really there: rather, they are brought into existence by an intricate interaction between light, crystals, and a sophisticated array of polarising filters. I was fortunate enough to study a howardite thin section for my undergraduate MEarthSci project at the University of Manchester.

3. ~ 380 howardites, ~ 1,400 eucrites, and ~ 500 diogenites.

Chapter 5: Cosmic Sediments

1. In addition to the CV chondrites, there are seven other groups within the carbonaceous class: CI, CM, CO, CR, CH, CB, and CK chondrites. Each group is classified on the basis of their distinct geological, chemical, and isotopic character, and probably originated from their own distinct asteroid. Carbonaceous chondrite groups are named after a prominent meteorite within that group. For example, the CO chondrites are named after a meteorite named Ornans: 'carbonaceous Ornans-like'.

2. McKeegan et al. (2011), Science, vol. 332, pp. 1528–32.

3. Patterson (1956), Geochimica et Cosmochimica Acta, vol. 10, pp. 230–7.

4. A notable early attempt to calculate the age of the Earth relied on religious texts. Bishop Ussher, a church leader in Northern Ireland in the seventeenth century, calculated the date of Creation in his 1650 publication Annales veteris testamenti (Annals of the Old Testament) by adding up the timescale over which biblical events unfolded. Relying on a literal interpretation of the Bible, he concluded that the Earth was created in 4004 BC (specifically, on the 23rd of October that year). That's precision geologists can only dream of, but of course, it is no use being precise if your numbers are wildly inaccurate. While Ussher displayed an impressive feat of biblical scholarship, Patterson proved he was wrong by a factor of one million.

5. Amelin *et al.* (2010), *Earth and Planetary Science Letters*, vol. 300, pp. 343–50.

Chapter 6: Drops of Fiery Rain

1. Sorby (1877), *Nature*, vol. 15, pp. 495–8.
2. Many thousands of chondrules have been described and analysed in minute detail across the scientific literature, but comparatively few have ever been dated using the lead clock. Dating chondrules using this method is exceedingly difficult, and only a small handful of laboratories worldwide are equipped to do it, including the laboratory I work in at the British Geological Survey (Nottingham, UK). I am fortunate enough to have personally dated a chondrule (it was 4,564 million years old, give or take a million years either side).

Chapter 7: Stars Down a Microscope

1. Burbidge, Burbidge, Fowler, and Hoyle (1957), *Reviews of Modern Physics*, vol. 28, pp. 547–650. This scientific paper, often abbreviated B^2FH in reference to the authors' surnames, is one of the most important and influential pieces of scientific work ever published. It clearly and elegantly described a coherent explanation for the origin of the chemical elements inside stars.
2. Abbott *et al.* (2017), *Astrophysical Journal*, vol. 848, L2. This paper has 3,664 authors. It is a true exemplar of the collaborative and international nature of modern science.
3. Heck *et al.* (2020), *Proceedings of the National Academy of Sciences*, vol. 117, pp. 1884–9.

Chapter 8: Star-tar

1. Anders *et al.* (1964), *Science*, vol. 146, pp. 1157–61.
2. Hamilton (1965), *Nature*, vol. 205, pp. 284–5.

3. Sagan and Khare (1979), *Nature*, vol. 277, pp. 102–7.

Chapter 9: Pieces of the Red Planet

1. We can see five planets in the night sky with the naked eye. The sixth is always visible: we live on it.
2. McKay *et al.* (1996), *Science*, vol. 273, pp. 924–30.

Chapter 10: Calamitous Tales

1. Alvarez *et al.* (1980), *Science*, vol. 208, pp. 1095–108.
2. Hildebrand *et al.* (1991), *Geology*, vol. 19, pp. 867–71.
3. DePalma *et al.* (2019), *Proceedings of the National Academy of Science*, vol. 116, pp. 8190–9.
4. Plutonium-239 is a synthetic isotope not at all present on the Earth, and is created inside nuclear reactors by bombarding uranium with neutrons. We humans have the power to do what stars do: create new elements. I find this simultaneously electrifying and terrifying.
5. Unsalan *et al.* (2020), *Meteoritics and Planetary Science*, vol. 55, pp. 886–94.

Index

absolute chronology 118–19
achondrites
  classification ix
  formation 81, 86, 89, 106
  iron *see* iron meteorites
  primordial lead 121–2, 123,
    125–6
  stony 44, 57–8, 97–105
  stony iron 44, 94–6
  see also pallasites
Africa 42–3, 222
Alabama, USA 259–60
Alais meteorite, France 191
alanine 201
Alberta, Canada 31
Aldsworth meteorite,
  Gloucestershire 279
Allan Hills 81005 50–1, 52, 218,
  227, 287–8
Allan Hills 84001 230–6
Allende meteorite, Mexico
  111–14, 128, 163, 175–6, 185
Allgemeine Naturgeschichte (Kant)
  65
aluminium-26 80, 85, 163
aluminium oxide 72

Alvarez, Luis 244–6, 249
Alvarez, Walter 244–6
Amelin, Yuri 128
amino acids 197–8, 199, 200–3
ammonia 73, 204
ammonites 243
angular momentum: conservation
  of 69–70
anorthite 50, 51
Antarctica 39–42, 49–50, 100, 110,
  133, 199n, 228, 230
Apennines, Italy 244
Apollo program 53, 112n, 185, 199,
  270
Appley Bridge meteorite,
  Lancashire 280
Argentina 10–12
argon 160, 174
aroma 44, 187, 191, 199n
art: prehistoric 1–2
Ashdon meteorite, Essex 281
Asteroid Belt
  discovery 25–7
  Kirkwood gaps 27–8, 32, 54
  number of asteroids 32, 79
  orbital resonance 28–9

Asteroid Belt (*continued*)
  proof of separate formation 49
  as source of meteorites 29,
    31–2, 48–9
asteroid deflection 267
asteroids
  appearance 55
  collisions 96, 102–3, 250–1,
    253–5, 256
  composition 55–6, 58
  energy release 250–1, 253
  exploration 54
  formation 58–9, 78–80, 106,
    109–10
  K–Pg extinction 245–7
  molten 86–9
  orbits 65
  organic molecules 207, 208,
    209–10
  as record of early Solar System
    history 81
  sunlight reflectance spectrum
    103–4
  temperature 58–9
astronomical unit 24
atomic clocks 118–21, 128, 232
atoms 45, 63–4
Australia 186–7, 198
Azaro, Frank 244–6

Banks, Joseph 20
Barthold, Charles 134
Barwell meteorite, Leicestershire
    258–9, 260, 282
basalt 51, 98, 100, 102, 221, 225–6
beads 12–13

Beddgelert meteorite, Gwynedd,
    Wales 281
Bennu 54, 271
beryllium 173n
Big Bang 66, 155–6
Bingöl province meteorite, Turkey
    97
bismuth 167
bituminous rocks 191
Blombos Cave, South Africa 1
Boisse, Adolphe 29, 89
boron 173n
Bovedy meteorite, County
    Londonderry, Northern
    Ireland 282
bridgmanite 92
Britain: meteorites x–xi, 277–84
Brown, Harrison 122–3

calcium 160
calcium-aluminium-rich
    inclusions (CAIs) 114–17,
    127–30, 145–6, 148, 162, 163
calcium oxides 72
Callisto 204
Cambridgeshire 283
Camel Donga meteorite, Australia
    38
Campo del Cielo meteorite,
    Argentina 10–12, 20–1, 38,
    90
Canada 31
Canyon Diablo meteorite, Arizona,
    USA 251–2
carbon
  abundance 188

isotopes 175, 177
nucleosynthesis 158
organic chemistry 188–9, 191, 203
carbonaceous chondrites
amino acids 200, 202–3
calcium-aluminium-rich inclusions (CAIs) 114–16, 129
chondrules 136, 141–5, 146–50, 162, 290
CI chondrites 192, 195, 197, 207–8
classification ix, 289
CM chondrites 196–7, 199–200, 202–3
composition 191–2, 195–200, 202–3, 207–8
CV chondrites 112–14, 129
formation 206–7
K–Pg extinction 264
name 113, 133
odour 187–8, 191, 198
carbonates 232
Cavaillé, Albert 194
cave art 1–2
Ceres 25, 26, 32, 54
Chaco people 10–11
chassignites 221, 222, 224, 227, 229–30
Chassigny meteorite, France 219–20, 224
Chelyabinsk, Russia 260–3
chemical analysis 21, 90
Chicxulub crater 247–9
chirality 200–2

Chladni, Ernst Florens Friedrich 16–18, 21, 23, 27, 30, 93–4
chondrites
carbonaceous see carbonaceous chondrites
classification ix, 110, 289
enstatite 110
formation 81, 109–10, 126, 127
noble gas isotopes 174–5
ordinary 133, 135–7, 141–5, 147, 150–1
chondrules 136, 141–5, 146–50, 162, 290
chromium 72n, 160, 263–4
chronology 118, 123
CI chondrites 192, 195, 197, 207–8
classification ix, 43–4, 81, 110, 113, 133, 289
clasts 50
climate change 270
clinopyroxene 222
Clinton, Bill 233–4
Cloëz, François Stanislas 190–1, 195
CM chondrites 196–7, 199–200, 202–3
coal 194
cobalt 167
Cold Bokkeveld meteorite, South Africa 191–2
Cold War 30–1
comets
composition 79
formation 79
meteor showers 273–6

comets (*continued*)
  orbits 25, 65
  organic molecules 205
condensation 72–3, 114, 115–16
conquistadors 10–11
conservation of angular
    momentum 69–70
cores of planets/planetesimals 87,
  88, 89, 92, 93
corundum 72
cosmic dust 38
cosmic rays 56–7, 173n, 204, 205
cosmochemistry 21, 180
cosmogenic nuclides 56–7
craters
  on asteroids 55, 102, 105
  on Earth 247–9, 251–2, 253–4
  on Moon 51, 53
creation myths 1, 63
Cretaceous–Paleogene mass
    extinction event 241–7
Croatia 15–16
Crumlin meteorite, County
    Antrim, Northern Ireland
    280
crystal formation
  condensation 72–3, 114, 115–16
  glass 51–2, 55
  howardites 99
  speed of cooling 51–2
  Widmanstätten pattern 90
Curiosity rover 236
CV chondrites 112, 114, 129
cyanogen 205
Czech Republic 31–2, 256
Czechoslovakia 30–1, 99–100, 133

dagger: ancient Egyptian 13–14
Danebury meteorite, Wiltshire
  282
dark flight 139
Dawn spacecraft 55, 105
deaths 220n, 260
deep time 3–4, 5, 117–18, 129–30
degradation 39
Denmark 244
deserts 39–43
Devonshire 277
diamonds 56, 176–7, 255
differentiation 86
dinosaur extinction 241–3, 247
diogenites 100–1, 102
Discoveries in the Theory of Sound
  (Chaldni) 16
dogs 220n

Earth
  age 59, 120–2, 126, 289
  asteroid collisions 250–1, 253–5,
    256
  atmosphere 57, 138–9
  core 89, 91, 245
  cosmic rays 57
  craters 247–9, 251–2, 253–5
  Cretaceous–Paleogene mass
    extinction event 241–7
  formation 76–7, 96, 145
  geological history 3–4, 5–6
  isotopes 46, 47–8
  mantle 92
  orbit 24
  size 38
  as source of life 63–4

earthquakes 91, 253–4
East Antarctic Ice Sheet 39–42,
    49–50, 100, 110, 133, 199n,
    228, 230
Efremovka meteorite, Kazakhstan
    129
Egypt 13–15, 220
Einstein, Albert 170
El Nakhla el Baharia, Egypt
    220
elements 155, 164–5, 180
Elephant Moraine 79001
    meteorite, Antarctica 228–9
Elliott, Rob 283
Enceladus 204
Ensisheim meteorite, France
    134–5
enstatite chondrites 110
erosion 39, 266
Essex 281
Eta Aquarids 275
etching 90
eucrites 99–102
Europa 204
evolution 4, 63, 181, 210, 264–5
extinction events 241–7, 264–5,
    266
eyewitness accounts
  Alabama, USA 259–60
  Alberta, Canada 31
  Aldsworth, Gloucestershire,
    England 279
  Allende, Mexico 111
  Appley Bridge, Lancashire,
    England 280
  Ashdon, Essex, England 281

Barwell, Leicestershire, England
    258–9, 282
Beddgelert, Gwynedd, Wales
    281
Bovedy, County Londonderry,
    Northern Ireland 282
Chassigny, France 219–20
Chelyabinsk, Russia 260–3
Crumlin, County Antrim,
    Northern Ireland 280
El Nakhla el Baharia, Egypt
    220
Ensisheim, France 134
Glatton, Cambridgeshire,
    England 283
Hatford, Oxfordshire, England
    277
High Possil, Glasgow, Scotland
    278
Hraščina, Croatia 15–16
importance of 37
Ivuna, Tanzania 192
Launton, Oxfordshire, England
    278
Middlesbrough, Yorkshire,
    England 140, 279
Montauban, France 190
Morávka, Czech Republic
    31–2
Murchison, Australia 186–7,
    198
Murray, Kentucky, USA 196–7
Nogata, Japan 133–4
Park Forest, Illinois, USA 31
Perth, Perthshire, Scotland 278
Pontllyfni, Gwynedd, Wales 281

eyewitness accounts (*continued*)
Příbram, Czechoslovakia 31
Pueblito de Allende, Mexico 111
Rowton, Shropshire, England
279
Sariçiçek, Turkey 97
Shergotty, India 220
Strathmore, Perthshire, Scotland
280
Stretchleigh, Devonshire,
England 277
Sylacauga, Alabama, USA
259–60
Wold Cottage, Yorkshire,
England 18, 19, 277–8

falls 37, 258
feldspar 73, 98, 100–1
Fermi, Enrico 257
Field of the Sky *see* Campo del
Cielo, Argentina
finds 38–9, 54
fireballs 9, 15, 17–18, 20, 30–1
fossils 210, 233–5
486958 Arrokoth 204
France 134–5, 189–90, 191,
219–20
fusion crust 138, 139–40

Ganymede 204
Geminids 276
General Theory of Relativity 170
Genesis spacecraft 115
geological time 3–4, 59, 117–18,
129–30, 269
geology 4–5, 265–6

Germany 16, 253–5
glass 51–2, 55, 98, 100, 103, 138,
257–8
Glatton meteorite,
Cambridgeshire 283
Glenrothes meteorite, Fife,
Scotland 283
Gloucestershire 279
gold 168, 172
Gorai, Masao 40
Göttingen, Germany 16
Gould, Benjamin Apthorp 24
graphite 177
gravitation
Newton's laws 28
planetesimal formation 74–5,
146
star system formation 65–6,
68–9, 156
gravitational focusing 75, 76
gravitational waves 170–1, 172
Greg, Robert Phillips 138
guardians of meteorites 135
Gwynedd, Wales 281
Halley's Comet 205
Hambleton meteorite, North
Yorkshire 284
Harding, Karl Ludwig 26
Hatford meteorite, Oxfordshire
277
Hawaii 98
Hayabusa2 mission 54
Heat Shield Rock meteorite, Mars
218, 238
HEDs/Vestan
classification ix

diogenites 100–1, 102
eucrites 99–102
howardites 97–9, 102–3
origins 101, 104–5, 229–30
helium 156, 157, 174
Herschel, Alexander Stewart 140
Herschel, William 23, 26
hibonite 72
hieroglyphs 14
High Possil meteorite, Glasgow,
Scotland 278
history 269
hoaxes 193–5
Hodges, Ann 259
Howard, Edward 20–2, 39, 43,
94, 110, 136
howardites 97–9, 101, 102–3
Hraščina, Croatia 15–16
humans 63–4, 180, 181, 266,
269–70
hydrogen 156, 157
hydrogen bombs 249–50

Iapetus 204
ice formation in Solar System 73
ice sheets 41
Iceland 98
igneous rocks 98–9, 144, 221
Illinois, USA 31
India 220
Indonesia 2
Innisfree meteorite, Alberta,
Canada 31
interference colours 288–9
International Space Station 267,
271

Io 226–7
Iran 12–13
Iraq 260
iridium 244–6, 247, 248
iron
ancient Egyptians 13–14
asteroids 87
Chaco people's usage 10–11
composition of meteoritic iron
13–14
formation by condensation
73
nucleosynthesis 160, 164
ordinary chondrites 135
sapphires 72n
siderophiles 87
slow-process nucleosynthesis
166–7
Tepe Sialk beads 12–13
iron meteorites
age of Earth 121, 123, 125–6
appearance 89–90
classification ix, 44
composition 90–1, 92
pallasites 95–6, 283
sources 93–4, 149
transit times 57–8
'iron from the sky' 14
Ironmasses (Chladni) 17, 27, 30,
93–4
isotopes
carbon 175
chondrules 144
chromium 263–4
cosmogenic nuclides 56–7
definition 44–5

isotopes (*continued*)
  magnesium 163
  neutrons 165–8
  noble gases 174–5, 176
  oxygen 46–9, 96, 115, 223
  radioisotopes 80, 85, 118–20
  slow-process nucleosynthesis
    166–7
  spectra 123, 126, 127
Italy 244
Itokawa 54
Ivuna meteorite, Tanzania 192

Japan 133–4
Japan Aerospace Exploration
    Agency (JAXA) 54
Johnson Space Center 50, 111,
    232
Juno 26
Jupiter
  effect on Asteroid Belt 28, 78–9
  formation 77
  gravitation 227
  name 214
  orbit 24
  organic molecules 203

K–Pg boundary layer 244–6, 256,
    263–4
K–Pg extinction 243–7
kamacite 90, 92
Kant, Immanuel 65
Kazakhstan 129
Khare, Bishun 206
King, Dr Elbert 'Bert' 111–12
Kirkwood, Daniel 27–8

Kirkwood gaps 27–8, 32, 54,
    105
knowledge acquisition 3
Krasnojarsk region, Siberia 94
krypton 174, 175

Lake House meteorite, Wiltshire
    22, 283
Lancashire 280
Laplace, Pierre-Simon 23, 51,
    65–6
Laser Interferometer Gravitational-
    Wave Observatory (LIGO)
    170
Launton, Oxfordshire 278
lead 119–20, 121, 122–5, 128
Leicestershire 258–9, 260, 282
Lennon asteroid 105
Leonids 276
Lichtenberg, Georg Christoph
    16–17, 18
life
  amino acids 197, 202
  carbon 189, 192
  Mars 232–6
  origins 195, 209–10, 265
light: speed of 24
light years 67n
LIGO (Laser Interferometer
    Gravitational-Wave
    Observatory) 170
lithophiles 88, 92, 121
Lonewolf Nunataks 94101
    meteorite, Antarctica 199n
Lovering, John 187–8, 198–9
Luna robotic landers 53

lunar meteorites ix, 50–4
Lyrids 275

McKay, David 232, 236
magma 44, 87–8, 98–9, 102, 221, 222
magnesium 159, 163
magnetic fields 57, 68, 90–1, 95–6
Manhattan Project 122, 257
mantles 92–3
Maros, Indonesia 2
Mars
  atmosphere 229
  formation 77
  geology 237
  images 217, 236
  meteorites on 218
  name 214
  orbit 24
Mars space probes 215–16
Martian meteorites
  Allan Hills 84001, Antarctica 230–6
  chassignites 221, 222, 224, 227, 229–30
  classification ix
  nakhlites 221, 222, 224, 227, 229–30
  shergottites 221–2, 224, 227, 228–30
mass spectrometers 123, 126, 127
matrix 116–17
melilite 72
melting 85
Mercury 24, 77, 79, 214, 225
Meteor Crater, Arizona 251–2

meteor showers 273–6
meteorites
  classification ix, 43–4, 81, 110, 113, 133, 289
  composition 13–14, 21, 29, 40, 44
  names 38, 113, 133, 287–8
  number collected 238, 258
  orbits 57–8
  as record of early Solar System history 81
  shape 139
  size 38
  see also specific types
Meteoritical Bulletin Database 287
meteoroids 273–4
meteors
  etymology 9
  speed 9–10, 18, 30, 138–9
  temperatures 10
methane 73, 203, 204
Mexico 110–12
Michel, Helen 244–6
Michel-Lévy, Mireille Christophe 114
'microfossils' 235
micrometres 50n
Middlesbrough meteorite, Yorkshire 140–1, 279
Mighei meteorite, Ukraine 196
Millbillillie meteorite, Australia 38
minerals
  formation 51, 55, 71–3, 88–9
  oxygen content 45
moldavites 256–7

Montauban, France 190, 193

Moon
  as actual source of some
    meteorites 50–4
  composition 51, 52
  as early theoretical source of
    meteorites 23–4
  formation 77–8
  motion 213
  rocks collected 53, 112–13, 185,
    199
  terrae and *maria* 51, 52–3
moons 77, 203–4
Morávka meteorite, Czech
  Republic 31–2
mountains: tallest in Solar System
  55
Murchison meteorite, Australia
  186–8, 198–200, 202–3
Murray meteorite, Kentucky, USA
  196–7
mythology 1, 10, 63, 214

Nakhla meteorite 220, 224
nakhlites 221, 222, 224, 227,
  229–30
names 38, 113, 133, 287–8
nanodiamonds 177
NASA
  Apollo program 53, 112n, 185,
    199, 270
  Curiosity rover 236
  Dawn spacecraft 53, 105
  Genesis spacecraft 115
  Johnson Space Center 50, 111,
    232

New Horizons spacecraft 204
Opportunity rover 218
OSIRIS-REx mission 54
NASA *cont'd*
  Psyche mission 93n
  Viking space probes 216–17,
    229
  Voyager 1 space probe 226
Natural History Museum, London
  12, 20–1, 43, 259
nebulae
  appearance 67–8
  calcium-aluminium-rich
    inclusions (CAIs) 114
  coalescence 68–9, 109
  organic molecules 205
  star formation 81, 156
  stellar explosions 148, 162
neon 159, 174
Neptune 77, 203
neutron stars 161, 169, 170–2
neutrons 45, 155–6, 165–8
New Horizons spacecraft
  204
Newton, Isaac 16
Newton's Law of Universal
  Gravitation 28
nickel 21, 160, 164n
nitrogen 57, 175
noble gases 174–5, 176, 228–9
Nogata-shi meteorite, Japan
  134
Nördlingen, Germany 253–5
Northern Ireland 280, 282
nuclear reactions 56–7, 69, 80,
  156, 291

nuclear weapons 122, 126, 246, 249–50, 257–8
nucleosynthesis 155, 157–60, 162, 173, 180–1
nuclides: cosmogenic 56–7
odour 44, 187, 191, 199n
oil 189
Olbers, Heinrich Wilhelm Matthias 25, 26
olivine 44, 73, 92, 95–6, 100, 142, 222
Opportunity rover 218
Opticks (Newton) 16
Optumpa 12
orbital resonance 28–9
orbits 24, 28–9, 31, 57–8, 64, 75
ordinary chondrites 133, 135–7, 141–5, 147, 150–1
organic molecules 73, 188–9, 191, 203–7, 209
  see also amino acids
Orgueil meteorite, France 190–1, 192–5, 197, 207
Orion Nebula 67, 81
Orionids 275
Ornans meteorite, France 289
orthopyroxene 92, 98–9, 100–1, 102, 230
OSIRIS-REx mission 54
Oxfordshire 277, 278
oxides: exotic 72, 115
oxygen
  abundance in rocks 45
  isotopes 46–9, 96, 115, 223
  lithophiles 88

nucleosynthesis 158
ordinary chondrites 135

Paleogene Period 243
Pallas 26, 32
Pallas, Peter Simon 94
Pallas Iron meteorite, Siberia 94–5
pallasites ix, 94–6, 283
Park Forest meteorite, Illinois, USA 31
Patterson, Clair Cameron 123–6, 145–6, 251–2
peridot 44, 73
perovskite 72
Perseids 274, 275
Perth meteorite, Perthshire, Scotland 278, 280
petrol 124, 125n
physical laws 28, 69–70
Piazzi, Giuseppe 24–5
plagioclase 98, 100
planetary meteorites classification ix
planetesimals
  aluminium-26 80
  cores 87, 88, 89, 92, 93
  formation 74–6, 81, 145, 146, 149, 178
  Jupiter's effect 78–9
  mantles 92–3
  molten 85–8
  organic molecules 207
planets
  composition 77
  core 88, 89, 92
  earliest in Universe 157

planets (*continued*)
  formation 65–6, 76–8, 145
  isotopes 48
  motion 214
  orbits 24, 28, 64–5, 76
plastic 270
platinum 168
Pleiades star cluster 67
Pluto 204
plutonium-239 257, 291
pollution by lead 124
Pontllyfni meteorite, Gwynedd,
  Wales 281
'potato radius' 75
pre-solar grains 179, 181
Příbram meteorite,
  Czechoslovakia 31, 133
protons 45, 56, 155–6
protoplanetary disc
  chondrules 143, 146–7, 148
  condensation 72–3, 113
  formation 69, 70–1
  organic molecules 206–7
  planetesimals formation 80,
    110, 148–9
  simulations 150
  supernovae influence 71
Psyche 93n
Pueblito de Allende meteorite,
  Mexico 111–12
pyroxene 98, 100, 101

Quadrantids 275
quartz 246, 247, 248, 255
quenched melt 51–2, 248,
  256

radioactivity 58, 167, 168, 171
radioisotopes 80, 85, 118–20,
  163
rapid-process nucleosynthesis
  164–5, 167–70, 172
regmaglypts 139, 218
relative chronology 118
relativity 170
Rheasilvia impact crater, Vesta 55,
  104, 105
Ries impact crater 253–5
ringwoodite 92
rocks
  bituminous 191
  chronology 118–19, 123
  formation by condensation 73
  human records 1–3
  igneous 98–9, 144
  melting 85
  sedimentary 109
  shock metamorphism 53
  see also crystal formation;
    geology
rotation: conservation of angular
  momentum 69–70
Rowton meteorite, Shropshire 279
Royal Society 20, 137
rubies 72
Russia 260–3
rust 39
Ryugu 54

Sagan, Carl 206, 235
Sahara 42–3
St George's Church, Nördlingen,
  Germany 253, 255

sand 246, 266
sandstone 137–8, 266
sapphires 72
sarcosine 200
Sariçiçek meteorite, Turkey 97
Saturn 77, 203, 214
science and responsibility
    271–2
Scotland 278, 280, 283
sedimentary rocks 109, 118, 136
seeds 193–4
seismicity 91
shatter cones 254
shergottites 221–2, 224, 227,
    228–30
Shergotty meteorite, India 220
Shipley, John v, 18–19
shock metamorphism 53, 98
shocked quartz 246, 247, 248,
    255
Shropshire 279
Siberia 94
siderophiles 87, 245
silicon 159
silicon carbide 177
silver 168
simulations 150
SJ101 (CAI) 128
slow-process nucleosynthesis
    164–7
smell 44, 187, 191, 199n
Smithsonian Institution 50
SNC clan 221
SNC fractionation line 223
snow line 73
solar flares 147–8

Solar System
    age 117, 127–9, 146
    formation 106, 127, 146–7,
        163–4, 173, 178, 206–7
    history 5, 63, 65–7, 69–78, 80,
        81
    orbits 24, 28, 64–5
    snow line 73
    tallest mountain 55
solar wind 115, 117
Sorby, Henry Clifton 137–8,
    141–3, 145, 147
South Africa 1, 191
Southern Delta Aquarids 275
space exploration 267–8, 270–1
spacetime 170
Spain 256
spectra
    asteroids 103–4
    iron meteorites 123
    neutron star collision 172
spinel 72
spinning: conservation of angular
    momentum 69–70
Stannern meteorite,
    Czechoslovakia 99–100
star-tar 206, 209, 235, 265
stardust 177–9, 180
stars
    effect on nebulae 68, 148
    formation 69, 70, 156–7
    mass 157
    motion 25, 213–14
    neutron stars 161, 169, 170–2
    nucleosynthesis 155, 157–60,
        162, 173, 180–1

stars (*continued*)
    rapid-process nucleosynthesis
      164–5, 167–70, 172
    slow-process nucleosynthesis
      164–7
    spectra 172
    stellar giants 70–1, 148
    supernovae 71, 148, 160–1, 169
static electricity 148–9
steepness ½ 47–8, 223
stellar winds 162, 163–4
Stevns Klint, Denmark 244
stony achondrites 44, 57–8,
    97–105
stony-iron achondrites 44, 94–6
    see also pallasites
Strathmore meteorite, Perthshire,
    Scotland 280
Stretchleigh meteorite, Devonshire
    277
strontium 172
suevite 248, 255
sulphur 159–60
Sun
    formation 69, 70
    future fate 158
    myths 1
    solar flares 147–8
    solar wind 115, 117
supernovae 71, 148, 160–1, 169
Sylacauga meteorite, Alabama,
    USA 259–60

taenite 92
Tanzania 192
tektites 255–8

Tepe Sialk, Iran 12–13
terrestrial fractionation line 47–8
thin sections
    Allan Hills 81005 50
    eucrites 100
    howardites 98
    interference colours 288–9
    ordinary chondrites 141–2
    pallasites 95
    Sorby's samples 137–8, 141–2
tholins 206
Thomson, William 288
timescales: geological 3–4, 59,
    117–18, 129–30, 269
Titan 203
titanium 72n, 160
Topham, Edward 19, 20, 22, 110
trade: illegal 42–3
trajectories 139
transit times 57–8
trinitite 257–8
Trinity nuclear bomb 257–8
Triton 204
Tsar Bomba 249–50
Turkey 97
Tutankhamun 13–14

Ukraine 196
Ultima Thule 204
Universe
    age 163
    Big Bang 66, 155–6
    history 156
uranium 118–20, 121, 122, 126,
    128, 146, 164, 257
Uranus 64n, 77, 203

Ursids 276
USA
  Alabama 259–60
  K–Pg boundary layer 256
  Meteor Crater, Arizona 251–2
  Murray, Kentucky 196–7
  Park Forest, Illinois 31
Ussher, Bishop: age of Earth
  calculation 289

Veneneia impact crater, Vesta
  105
Venus 24, 64n, 77, 214, 225–6
Vesta 26, 32, 55, 104, 230
Vestoids 105
Vigarano meteorite, Italy 112n,
  114
Viking landers 216–17, 229
Virgo Interferometer 170
volatile substances 116n
volcanic activity 101–2, 224–5,
  226
Voyager 1 space probe 226

wadsleyite 92
Wales 281

water
  ice formation in Solar System
    73
  Martian meteorites 223, 231–2
  meteorites 44, 191, 194, 207–8
  Solar System locations 203–4
Widmanstätten pattern 90, 92, 288
Wiltshire 22, 282, 283
Wöhler, Friedrich 192
Wold Cottage meteorite,
  Yorkshire v, 18–20, 22, 110,
  133, 277–8
writing: invention of 2–3, 269,
  271

xenon 174–5

Yamato 791197 meteorite,
  Antarctica 54
Yorkshire 18–20, 137–8, 140,
  277–8, 279, 284
Yorkshire Museum 141
Yoshida, Masaru 40
Yucatán Peninsula 247–9

zircon 123–4, 125

Tim Gregory was born and raised in West Yorkshire, and since starting his rock collection aged four he has held a continued fascination with the natural world. Following an undergraduate degree in geology and a PhD in cosmochemistry, Tim is now a research scientist at the University of Bristol and the British Geological Survey (Nottingham).

A geologist by training, Tim uses his knowledge of rocks to study stones that fall from the sky: meteorites. His research specifically focuses on the geological and chemical make-up of meteorites, and the timing of events that unfolded as our Solar System was forming 4.6 billion years ago. Tim is a science populariser who regularly speaks with live audiences, often features on radio, and occasionally presents on television. *Meteorite* is his first book. When not in the lab or writing, Tim is usually either hiking, photographing landscapes, or playing his guitar.